# DATA-DRIVEN SMART COMMUNITY DESIGN

This book couples data analytics with social behavioural studies and participatory design to derive deeper insights on city dwellers' present needs and future aspirations, thereby enabling the development of targeted spatial and programmatic interventions for diverse communities.

Public housing in Singapore has been regarded internationally as a success story. This book outlines the latest strategies and concepts for addressing the emerging social challenges of the ageing population: shrinking household size, increasingly diverse demographics and widening inequality, and fostering inclusive and resilient neighbourhoods. Adopting an interdisciplinary approach, this book:

- Outlines an innovative data-driven planning process for housing neighbourhood and community design
- Provides a framework for planners and designers to synthesise qualitative and quantitative data analyses
- Presents a comprehensive set of tested urban analytics tools, digital platforms and participatory toolkits used to design and develop community initiatives.

A recommended text for students undertaking urban planning, urban design, housing design, architecture, real estate, urban sociology and community design, the book's strategies for evidence-based neighbourhood designs will also appeal to practitioners and policymakers.

**Keng Hua Chong** is Provost of Nanyang Academy of Fine Arts at the University of the Arts Singapore. He is formerly Associate Professor of Architecture and Sustainable Design at the Singapore University of Technology and Design.

# DATA-DRIVEN SMART COMMUNITY DESIGN

## Strategies for Fostering Inclusive and Resilient Neighbourhoods

*Edited by Keng Hua Chong*

Routledge
Taylor & Francis Group

LONDON AND NEW YORK

Cover image: Natasha Yeo Min

First published 2025
by Routledge
4 Park Square, Milton Park, Abingdon, Oxon, OX14 4RN

and by Routledge
605 Third Avenue, New York, NY 10158

*Routledge is an imprint of the Taylor & Francis Group, an informa business*

*British Library Cataloguing-in-Publication Data*
A catalogue record for this book is available from the British Library

*Library of Congress Cataloging-in-Publication Data*
Names: Chong, Keng Hua, 1977– editor.
Title: Data-driven smart community design: strategies for fostering inclusive and resilient neighbourhood / edited by Keng Hua Chong.
Description: Abingdon, Oxon; New York, NY: Routledge, 2025. |
Includes bibliographical references and index.
Subjects: LCSH: City planning—Statistical methods. | Sociology, Urban. |
Smart cities. | Neighborhoods.
Classification: LCC HT166 .D347 2025 | DDC 307.1/416—dc23/eng/20240912
LC record available at https://lccn.loc.gov/2024029939

ISBN: 978-1-032-56922-2 (hbk)
ISBN: 978-1-032-56921-5 (pbk)
ISBN: 978-1-003-43765-9 (ebk)

DOI: 10.4324/9781003437659

Typeset in Times New Roman
by codeMantra

# CONTENTS

# CONTRIBUTORS

## Editor

**Keng Hua Chong** is Notes on Associate Professor in the Architecture and Sustainable Design Pillar at Singapore University of Technology and Design, where he directed the Social Urban Lab (SOULab) and the Positive City Lab and led the MND-HDB New Urban Kampung Research Programme. He is also a Founding Partner of COLOURS: Collectively Ours, an award-winning design practice specialising in public space and social impact.

## Contributors

**Malik Mohamed Barakathullah** is Research Fellow I in the Architecture and Sustainable Design Pillar at Singapore University of Technology and Design.

**Roland Bouffanais** is a Complexity Scientist and an Associate Professor in the Department of Computer Science & Global Studies Institute at the University of Geneva, Switzerland.

**Sin Mei Cheng** is Research Associate in the Social Urban Lab (SOULab) at Singapore University of Technology and Design.

**Mihye Cho** is Associate Professor of Sociology in the Department of Sociology at SungKongHoe University, South Korea.

**Sofia Foo Xin Di** is Business Development Executive at SAA Architects, Singapore. She was formerly Research Associate in the Social Urban Lab (SOULab) at Singapore University of Technology and Design.

**Pia Fricker** is Associate Professor of Augmented Computational Design, Vice Head of Department of Architecture, School of Arts, Design and Architecture, and Head of Urban Studies and Planning Major in Architecture at the Aalto University, Finland.

**Tshui Mum Ha** is a PhD student in the Department of Geography at the Ohio State University. She was formerly Research Associate in the Social Urban Lab (SOU-Lab) at Singapore University of Technology and Design and Project Coordinator of MND-HDB New Urban Kampung Research Programme.

**Yohei Kato** is Research Fellow in Lee Kuan Yew Centre for Innovative Cities at Singapore University of Technology and Design.

**Zi-Yu Khoo** is a PhD candidate in the School of Computing at National University of Singapore. She was formerly Research Engineer at EDF Lab Singapore.

**Jiyoun Kim** is Senior Researcher at Hansung University, South Korea.

**Zann Koh** is Research Fellow I in the Engineering Product Development Pillar at Singapore University of Technology and Design.

**Billy Pik Lik Lau** is Research Fellow I in Temasek Labs at Singapore University of Technology and Design.

**Denise Nicole Lim** is Architectural Designer I in KPF, Singapore. She was formerly Senior Research Assistant in the Social Urban Lab (SOULab) at Singapore University of Technology and Design.

**Ran Liu** is Senior Research Fellow in the Engineering Product Development Pillar at Singapore University of Technology and Design.

**Hasala Marakkalage** is Research Scientist in NEXT Open Innovation, NCS Group, Singapore.

**Natasha Yeo Min** is Managing Partner of COLOURS: Collectively Ours, Singapore. She was formerly Senior Research Assistant in the Social Urban Lab (SOU-Lab) at Singapore University of Technology and Design.

**Sze Min Neo** is a manager in the Department of Project Development at Temasek Shophouse, Singapore. She was formerly Senior Research Assistant in the Social Urban Lab (SOULab) at Singapore University of Technology and Design.

**Aerilynn Tan** was formerly Senior Research Assistant in the Social Urban Lab (SOULab) at Singapore University of Technology and Design.

**Bige Tunçer** is Professor and Chair of Information Systems in the Built Environment at Eindhoven University of Technology, The Netherlands.

**Christine Yogiaman** is Founding Partner of Yogiaman Tracy. She was formerly Assistant Professor in the Architecture and Sustainable Design Pillar at Singapore University of Technology and Design.

**W. Quin Yow** is Professor of Psychology, Head of Humanities, Arts & Social Sciences Cluster, and Director of Language and Social Cognition Lab at Singapore University of Technology and Design.

**Chau Yuen** is Associate Professor in the School of Electrical & Electronic Engineering and Assistant Dean (International Engagement), Graduate College at Nanyang Technological University, Singapore.

**Shuang Zheng** is a PhD Candidate in the Humanities, Arts & Social Sciences Cluster at Singapore University of Technology and Design.

**Yuren Zhou** is Research Fellow I in the Engineering Product Development Pillar at Singapore University of Technology and Design.

# FOREWORD

The public housing developed by the Housing & Development Board (HDB) in Singapore is unique. It houses some 80% of Singapore's resident population and has brought home ownership to some 94% of its households, enabling Singapore to achieve one of the highest home ownership rates in the world. But public housing in Singapore is not only about housing. Through innovative policies, it undergirds many aspects of the social and lived experiences of its residents. At an individual level, it provides housing that caters to lifecycle housing needs of different segments such as young couples, families, the elderly and singles. The HDB home is also one of the key pillars for attaining retirement adequacy. At a broader level, HDB's design and allocation policies encourage ethnic integration and interaction that strengthens community and neighbourly ties. In 2017, as CEO of HDB, I was acutely aware that even as we continue to upgrade and build new towns, it was essential that we understand the changing complexion, aspirations and needs of our residents. We therefore initiated a grant call for research that would identify potential new resident 'archetypes' and what they perceive to be important factors that will meet their aspirations for a 'quality of life'. Such research insights, enabled by technology and data science, would help to inform on how we can improve the physical and social design of HDB towns.

The SUTD team stood out in proposing a ground-breaking interdisciplinary research programme that addressed HDB's intent. This research programme, supported by the Ministry of National Development and the National Research Foundation under the original title 'New Urban Kampung', sought to introduce innovations to urban planning by leveraging the power of data analytics and social behavioural studies for evidence-based interventions and to foster inclusive and resilient communities. I had the pleasure of being involved in several discussions with the SUTD team as the research progressed and interesting insights unfolded.

The years encompassing this research were not without challenges, particularly during the unprecedented upheaval caused by the COVID-19 pandemic. Faced with the constraints of physical distancing and community engagement, the team showed its resilience in pivoting swiftly, by deploying digital tools and innovative approaches to engage the community and to ensure the continuity and success of the project.

Spanning six years of research, documentation and reflection, this book is the culmination of the research team's collective efforts to redefine urban living and community design in Singapore. Through a comprehensive exploration of three distinct towns – Toa Payoh, Jurong East and Punggol – the team has uncovered insights that could potentially be scaled and applied to diverse urban contexts.

Divided into three parts, the book elucidates a clear and systematic process of data-driven design. Part I explores the intricacies of resident segmentation and quality of life assessment, employing innovative methodologies to understand the diverse needs and perspectives of our communities. Through multi-modal data collection and analysis, useful insights were gleaned on the factors shaping residents' everyday experiences and aspirations.

Part II showcases the development of cutting-edge urban analytics tools and processes aimed at identifying opportunities for targeted interventions and informed decision-making. These tools can empower planners and designers to optimise resource allocation and enhance community well-being.

In Part III, attention was turned to community enablement and inclusivity, highlighting the pivotal role of participatory research in driving meaningful change. The Community Enablement Playbook offers a suite of strategies for co-learning, co-creation and co-management tailored to the unique needs of each neighbourhood.

I am confident that the insights gleaned from this research will provide urban planners, policymakers and community leaders with fresh perspectives. May this book ignite new conversations, spark innovative solutions and pave the way for a brighter, more equitable future for all.

**Professor Cheong Koon Hean**
Chair, Lee Kuan Yew Centre for Innovative Cities,
Singapore University of Technology and Design
Chair, Centre for Liveable Cities Advisory Panel,
Ministry of National Development, Singapore
Former CEO, Housing & Development Board
and the Urban Redevelopment Authority

# PREFACE & ACKNOWLEDGEMENTS

This book is the culmination of six years of research, application, documentation and reflection, spanning from 2017 to 2023, and represents the collective efforts of numerous individuals dedicated to the exploration of new urban communities empowered by data analytics, social behavioural studies and participatory design. Originally titled 'New Urban Kampung', this interdisciplinary research programme engaged scores of researchers and policymakers, navigating through the challenging yet transformative period of the COVID-19 pandemic. This period demonstrated the significance of technology and empathic design in fostering connections and strengthening communities.

This research program was indeed ambitious, consisting of four distinct yet intricately interlinked projects conducted simultaneously. I was honoured to be appointed as the Program Lead, fully aware of the significant challenge it posed. Fortunately, I had strong support from highly knowledgeable, talented, and dedicated colleagues from diverse fields, including psychology, sociology, geography, data science, engineering, architecture, urban planning, and community design, to name a few. More importantly, this research built upon our close working relationships from previous projects: *Creative Ageing Cities* (co-led with Dr. Mihye Cho), *Living in an Ageing World: New Elderly Housing and Neighbourhood Design Typology* (co-led with Dr. W. Quin Yow), *Dande-lier @ i Light Marina Bay* (co-led with Dr. Chau Yuen), and *Liveable Places: A Building Environment Modelling Approach for Dynamic Place-making* (led by Dr. Bige Tunçer). These collaborations provided us with a deep understanding of each other's expertise and working styles, which greatly facilitated the success of this unprecedented large-scale interdisciplinary research program. On a personal note, the publication of this book marks a significant milestone in my thirteen years with the Singapore University of Technology and Design. I am immensely thankful for all the opportunities and friendships I have had over these past 13 years.

On behalf of the research team, I would like to extend our heartfelt gratitude to the following organisations and individuals for their invaluable contributions to the success of this endeavour:

Singapore University of Technology and Design (SUTD), for providing essential research infrastructure and administrative support throughout the programme.
Housing & Development Board (HDB), for spearheading and co-investigating the New Urban Kampung Research Programme, with special acknowledgement to Co-Investigator Ms. Tan Gee, Director (Centre of Building Research), Building Research Institute, HDB, for her leadership and collaboration.
Urban Redevelopment Authority (URA), for their collaborative partnership and support in advancing the research objectives.
EDF Lab Singapore, for their collaboration in the development of the City Application Visualisation Interface (CAVI), a crucial tool integrating urban analytic models developed in this research programme.
Ministry of National Development (MND) and Centre for Liveable Cities (CLC), for their support in publishing articles and organising forums to disseminate the knowledge gained from this research.
Aalto University, Finland, for supporting my research stint in 2023–2024 and enabling the research collaboration and cross-cultural knowledge exchange between the two universities.

The research team is deeply grateful to the following experts for their guidance and expertise:

Professor Chan Heng Chee, Ambassador-at-Large, Singapore Ministry of Foreign Affairs, and Former Chair, SUTD Lee Kuan Yew Centre for Innovative Cities.
Professor Cheong Koon Hean, Chair, SUTD Lee Kuan Yew Centre for Innovative Cities; Chair, Centre for Liveable Cities Advisory Panel, MND, Singapore; Former CEO, Housing & Development Board and the Urban Redevelopment Authority.
Professor Erwin Viray, SUTD Chief Sustainability Officer and Former Head of Pillar, SUTD Architecture and Sustainable Design.
Er. Dr. Johnny Wong Liang Heng, Deputy CEO (Building), Housing & Development Board.

Finally, special recognition goes to Ms. Ha Tshui Mum for her dedication as Project Coordinator of the New Urban Kampung Research Programme at SUTD, Mr. Jeremy Chang for his continuous support in facilitating communication between HDB and SUTD, and Ms. Cheng Sin Mei for her invaluable assistance during the editorial process of this book.

This research, led together with the Housing & Development Board, was supported by the Singapore Ministry of National Development and the National

Research Foundation, Prime Minister's Office under the Land and Liveability National Innovation Challenge (L2 NIC) Research Programme (L2 NIC Award No. L2NICTDF1-2017-4). Any opinions, findings and conclusions or recommendations expressed in this material are those of the author(s) and do not reflect the views of the Housing & Development Board, Singapore Ministry of National Development and National Research Foundation, Prime Minister's Office, Singapore.

**Dr. Chong Keng Hua**
Provost, Nanyang Academy of Fine Arts,
University of the Arts Singapore
Former Associate Professor, Architecture and Sustainable Design,
Singapore University of Technology and Design
Programme Lead, New Urban Kampung Research Programme

# INTRODUCTION

*Keng Hua Chong*

## Little Library Movement

Towards the later stage of the COVID-19 pandemic, as physical movements and social interactions gradually became less restricted, there was an interesting ground-up movement initiated by residents in various housing estates of Singapore – the Little Library movement.

It started with a resident from Ghim Moh estate, a young full-time mother, who put together some 'rescued' books, a wooden crate to hold the books, a chair left by her friend and a plant from her living room at the common lift landing on the 39th floor where she stayed. She labelled the small nook "Little Library", hoping to attract young children to start picking up the books and hence the habit of reading, and to get to know other "book-loving families" at her place. The newly set-up lift lobby looked comfortably cosy with abundance of natural light; no wonder it became an instant hit with the neighbours. Through her circle of young mothers in the neighbourhood, another group that she started, and later through open interest groups on social media, the Little Library movement took off (CNA, 2021, February 21).

Little Libraries sprang up around the island, from void decks (empty ground-floor space of public housing blocks) at Yishun in the north and Boon Lay in the west, to a clinic at Tampines in the east (The Straits Times, 2022, March 19, 2023, November 6). One of the more extensive ones was at Holland Village. Inspired by the Ghim Moh story, a senior music teacher populated a corner of her block with 11 bookshelves and more than 5,000 books, along with a table tennis table, a few guitars and a piano – all donated by neighbours and visitors. The housing block's void deck had then become a regular hangout place among neighbours (The Straits Times, 2023, January 30).

DOI: 10.4324/9781003437659-1

However, the process of making this happen was not without hiccups, as books and musical instruments were sometimes stolen. Another issue that surfaced was the dumping of unwanted things, including expired food, by inconsiderate residents. Getting the other residents to co-manage the place that opened round the clock was also a challenge.

Among these Little Libraries, the one at Boon Lay seemed to have the worst fate. Two bookshelves were stolen just hours after the library was set up by the founder. In the following days, all the books were gone as well, only to magically reappear again after the story went on national news (The Straits Times, 2023, April 23). The founder persevered at first, but lost hope again when the place was left in disarray by a child climbing all over the shelves and scattering the books all over the floor. That was it. It was too much for him alone to handle, and he decided to close the Little Library for good. A social media post likened this to a "well-written Greek tragedy, bringing us full circle to where we started" (The Straits Times, 2023, October 20).

## New Community Dimensions in Housing

The Little Library was one of a handful of community ground-up movements that took place in Singapore's housing estates in recent years. Some were organically started by residents; some were seeded and supported by the government agencies. In a city-state that has been well-known for its "highly centralised approach to urban development" with "little space for civic engagement", the emerging activities on the ground indicate a shift in landscape of urban development – it has gradually become more open, and citizens are now encouraged to get involved (Cho & Kriznik, 2017).

This is, however, not to say that community development was not the goal in Singapore's earlier urbanisation efforts. In fact, since the Housing & Development Board (HDB) was set up in 1960, having to relocate large population from their 'kampung' (village in Malay) houses into new modern housing blocks which had inevitably ruptured the original neighbourly ties and kinships, it had aimed to rebuild communities in the new estates, especially in the effort of ethnic integration through flat allocation (Fernandez, 2011). This process corresponded to Singapore's post-independent nation-building effort, with the idea of constructing a homogeneous public housing social landscape with mixed-age, mixed-ethnicity, and mixed-income neighbourhoods – a "showcase of a middle-class society" (Tan, 2017), while reinforcing a newly created national identity that is "regardless of race, language or religion".

Yet, heterogeneity in public housing neighbourhoods has occurred naturally over the past six decades due to self-selection processes by people and ramping-up of immigration in recent years (Tan, 2017). Diversity was observed in economic class, ethnic and nationality, with a trend of elitism based on affiliation to schools, workplaces and voluntary associations, which scholars have identified as potential sources of social boundary or even tension and conflict (Tan, 2017; Chua et al., 2020). External influences such as geo-political conflicts added to the divide in opinions and political views. Hence, community building in housing neighbourhoods has become even more crucial and urgent than ever, but the way that it is done would not be the same as before.

In an increasingly complex and uncertain world, housing development needs to go beyond what was done in the past, i.e. merely constructing mixed-age, mixed-ethnicity and mixed-income neighbourhoods. As society matured, residents "would increasingly see themselves less as customers to be served, but more as empowered stakeholders with a responsibility to serve the community" (Tan, 2017). Therefore, the new model needs to:

1 leverage on the *diversity* of the residents and their social capital;
2 ensure *equal access* to opportunities and good Quality of Life (QoL) such that the different groups are attended to fairly in neighbourhood planning and design; and
3 build an *inclusive process* that enables residents to collaborate in creating and developing communities in response to their needs.

Such collective values of diversity, equity and inclusion (DEI) have been adopted by many organisations recently. In fact, they are rooted in the 1960s when the community design movement took off. Community design stands for an alternative to professional planning practice that emphasises "the involvement of local people in the social and physical development of the environment in which they live" (Sanoff, 2006). The vision is to see "government participation in citizen initiative", instead of the conventional idea of citizens participating in government programmes (Sanoff, 2006). It is with this in mind that we embarked on the study of enhancing the practice of community design through leveraging new technology and data-driven processes, to test the hypotheses of enabling more diverse communities, ensuring more equal access and better QoL and rethinking inclusive participation through emerging platforms. This is the vision of "Smart Community Design" proposed in this book.

Singapore public housing is adopted as the main subject of interest throughout the study as it is at a crossroad. There are more than 1 million flats in 23 towns and 3 estates in Singapore. Having successfully provided affordable and quality housing to about 80% of the total population with 97% of resident households owning their flats and 77% of residents living in larger flat types (Tan, 2017), and, more recently, having pioneered a range of environmentally sustainable features (Housing & Development Board [HDB], 2020), it is time to start looking into strategies towards social sustainability for the next phase of its development.

Moreover, its strong governance, high accountability and centralised planning process offer an interesting case to investigate the mechanisms needed to shift towards a more bottom-up approach to encourage, mobilise and enable the community. It is not so much to reverse the planning process, but to strike a balance and offer possible alternatives that provide a more open platform for citizen participation in planning and upkeeping while not losing efficiency and accountability.

Finally, Singapore has a population that enjoys a high level of education and occupational attainment. It also has one of the highest internet access rates. It is the world's fifth smartest city in 2024, top in Asia, ahead of Beijing, Taipei and Seoul (The Straits Times, 2024, April 12). In a digital age such context offers an opportunity to investigate the advantage of having high data proliferation, as these demographic characteristics could potentially be translated into social capital. But as we can see from the earlier example of the Little Library movement, that translation is far from straightforward.

What made some ground-up community projects succeed or fail, despite similar ideas? How can we understand the diversity and trend in the residents' demography, their relationship with the built environment and enable them accordingly? How can planners and architects incorporate these considerations in the already highly evolved practice of neighbourhood planning and design? How do we move forward in planning for and with future communities, leveraging technology and data? These questions thus form the premises for our research on a new frontier of data-driven community design in the context of residential neighbourhoods.

## Singapore's Public Housing and Community Development

The building of community and the planning and development of Singapore's public housing have always been interwoven.

The Central Area (city centre of Singapore) was used as experimental ground for new public housing typologies in the beginning, as there was an urgent need to resettle the slum population to carve out valuable lands for commercial and civic development to rejuvenate the city (Fernandez, 2011). The conscious rebuilding of community was done through its urban design to encourage more social interactions among residents, by developing pedestrian-friendly estates, incorporating commercial and recreational amenities in the lower floors and introducing public spaces within the housing complexes in the form of internal courtyards, landscaped plazas and walkable streets (Heng & Chong, 2006; Goh & Heng, 2017).

Some of these typologies were subsequently adopted in newly planned 'satellite towns' across the island, with social amenities and public spaces that were meant to "socialise residents" to modern high-rise living. Together with the 'Community Centres' run by the government's social arm People's Association (PA) and the formation of 'Residents' Committees' (RCs) that delegated official roles to grassroots leaders, there was a conscious and consistent drive from the government to reconstruct community for the newly relocated residents (Ooi & Tan, 1992; Yeoh & Kong, 1995). The formulation of the 'precinct concept' further reinforced the arbitrary construct of community, with its design evolution, along with the housing block's ever-increasing density and height, representing the changing concepts of community and the underlying national interests over the years (Chong & Tan, 2015). The evolution of housing planning ideologies in relation to community building efforts can be captured distinctively in the development of three towns – Toa Payoh, Jurong East and Punggol.

## Toa Payoh

Constructed in 1965–1977, Toa Payoh was HDB's first satellite town that was envisioned as a "self-sufficient community" (Fernandez, 2011). Based on the "neighbourhood principle", several neighbourhoods ringed around a town centre in a concentric manner, each with its own neighbourhood centre, community facilities and schools (Cheong, 2017). Housing blocks were designed with varying heights from 6 to 25 stories, with taller blocks serving as landmarks that signified the nation's high-rise strategy. Its town centre was the first "pedestrianised hub" (Fernandez, 2011) with low-rise slab blocks of shops and facilities that ranged from eateries, department stores, theatres, library, clinics and community institutions, connected to the largest bus terminal at the time. Recreational, sporting amenities and industrial areas sat on the fringe, bringing work and play close to where people live. The planning of Toa Payoh formed the basis of the New Town Structural Model (nicknamed "checkerboard estate"), which became the model for future new towns.

## Jurong East

The "precinct concept" was established in the 1980s to provide "more conducive social setting for neighbourly interaction and community bonding" (Cheong, 2017). The neighbourhoods were further broken down into several smaller housing precincts, each comprising about 400–600 dwelling units (HDB, 1995, 2005). A variety of recreational facilities like a precinct pavilion, playground, fitness corners, multi-purpose courts, soft and hard landscape, etc., were located at the precinct centre. Jurong East is one of the examples of such precinct planning. Housing blocks were designed with shorter common corridors shared by fewer units on each floor as an attempt to foster neighbourliness (Cheong, 2017). Another interesting observation is that the precinct planning was also loosely associated with the boundaries served by respective RCs, such that community welfare and event organisation are taken care of by voluntary grassroots leaders who are residents from the very precinct. Jurong East has since been transformed from an industrial town to a vibrant and liveable housing estate and commercial centre, especially in the recent years as it was envisioned as Singapore's second Central Business District (CBD), with its town centre extensively rejuvenated into the new Jurong Gateway Precinct as a model for future regional centre (Urban Redevelopment Authority [URA] website, accessed 2024, April 22).

*Punggol*

Shaped by the "Punggol 21" vision introduced in 1996, Punggol was designed as the first 'Eco-town' and "A Waterfront Town of the 21st Century" (Fernandez, 2011). It ushered in the latest phase of housing development emphasising on sustainability, prototyping eco-features that were subsequently incorporated in all public housing (Cheong, 2017). This period also saw the introduction of Community Development Councils (CDCs), where political leaders and community groups took over the running of various community assistance programmes, to foster community ties alongside housing provision (Fernandez, 2011). With "Punggol 21 Plus" announced in 2007, a new waterway was introduced as a key green and blue feature together with other plans to further enhance Punggol as a 'living laboratory' for sustainable development (SD), both environmentally and socially. They include more signature waterfront housing districts, more green parks and corridors that are integrated with waterfront promenade, more civic and recreational spaces such as high-rise gardens and community living rooms, to develop it into a "Community-Centric Town" – one of the three main thrusts of the HDB Roadmap to Better Living that guides its future development (HDB, 2020).

Other than developing new towns, estate renewal efforts in existing towns were considered equally important in strengthening community. These programmes include Enhancement for Active Seniors (EASE) to install age-friendly features for ageing residents and the Neighbourhood Renewal Program (NRP) that focuses on precinct- and block-level improvements to renew and provide a wider range of common facilities (Fernandez, 2011). A more extensive initiative Remaking Our Heartland (ROH) was launched in 2007 to renew the HDB 'heartland' in a more comprehensive way. It was under this programme that HDB estates were classified into "Matured Estates" (flats largely built before 1980s), "Middle-Age Estates" (flats largely built in 1980s) and "Young Estates" (flats built in 1990s and later) (Cheong, 2017). Toa Payoh, Jurong East and Punggol are generally representative of these classifications, respectively, and were selected to undergo the ROH programme.[1] Likewise, based on their representativeness, these three towns were also chosen as study sites in our research (Figure I.1).

## Enhancing Social Sustainability through Smart Community Design

An SD framework (Figure I.2) has been proposed by HDB, which sets out ten key desired sustainability outcomes to guide future public housing development. Among them six were environmental outcomes with wide-ranging environmental sustainability strategies and specific initiatives to reduce carbon emissions, optimise the use of resources and achieve effective water and waste management. Two outcomes were devoted to social sustainability – 'Meeting Society's Needs' and 'An Endearing Home' (Cheong, 2017; HDB, 2020).

**Study sites**

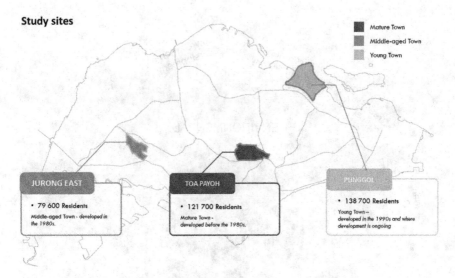

■ Mature Town
■ Middle-aged Town
■ Young Town

**JURONG EAST**
- 79 600 Residents

Middle-aged Town - developed in the 1980s.

**TOA PAYOH**
- 121 700 Residents

Mature Town - developed before the 1980s.

**PUNGGOL**
- 138 700 Residents

Young Town – developed in the 1990s and where development is ongoing

**FIGURE I.1**    Map of Singapore showing the locations and statistics of three selected towns.

*Source*: Authors.

Under social sustainability, more research and evidence were needed in the next phase to develop and substantiate strategies to promote stronger social integration, a sense of ownership and place identity among residents. Creating 'community-centric towns' is crucial for Singapore, to ensure that its multi-racial, multi-cultural society living in high-rise, high-density housing can live and work in harmony, especially when it is facing an era of unprecedented demographic transition.

Shrinking household size and evolving resident demography are already evident. According to Census 2020, the average household size decreased over the past decade from 3.5 persons to 3.2 persons across most housing types. Overall, about 60% of households had three or fewer members. During the same period, the percentage of single-person households increased 4.2% (from 12.2% to 16%), while that of couples with children decreased 8.3% (from 56.0% to 47.7%).

An ageing population adds to the challenge. About a third of households had at least one senior aged 65 years and above (increased from 24.1% to 34.5%), while senior-only households also doubled (from 4.6% to 9.3%) in the past decade. Singapore is set to attain "super-aged" status in 2026 when one in five citizens will be aged 65 and above. By 2030, the proportion will be one in four.

Against this backdrop of evolving household demography exist widening inequalities, increasingly diverse communities and their different perspectives and priorities. Traditionally, residents have been profiled based on demographic markers such as age, gender or ethnicity. But such stratifications have become inadequate to understand their circumstances and choices or to assess their QoL, let alone derive targeted approaches to support their complex needs.

**FIGURE I.2** Sustainable development framework, outlining the Sustainable Singapore Blueprint & Concept Plan, as well as Environmental Governance & Leadership. Image adapted from Housing & Development Board, Singapore: Home, Truly: Building Dreams, Housing Hopes, 2020, p. 30.

Furthermore, there is also the challenge of assessing the state of societal progress and the degree to which it enhances individual's QoL and well-being. The notion of QoL is fundamentally about a society's aspirations towards a 'good' life, and how to organise social and economic resources towards such goals. While there are existing QoL indexes that are potentially useful in facilitating inter-city comparison, they may overlook the highly localised and contextual social dynamics, as well as subjective and intangible drivers of well-being. Such contextualised perspective is critical when it comes to public housing neighbourhood, especially if we want to promote community bonding and participation.

As reported in the HDB Sample Household Survey 2018, while it is encouraging to see that most residents (97.1%) agreed that there were sufficient places for neighbourly interactions (mainly exchanging greetings through chance encounters), the neighbourly interaction had generally declined. It is noticed that participation in the community had declined substantially (from 45.4% in 2013 to 34.4% in 2018), mainly due to "lack of time". The trend of decreasing community participation when more community support is needed with a rapidly ageing population is indeed worth concerning.

Therefore, it is time to go beyond building physical infrastructure or 'hardware', and to put more efforts and resources in developing soft infrastructure or 'heartware' that would promote neighbourliness, expand residents' social network and provide new opportunities for community participation. These are the avenues that urgently need an evidence-based framework, appropriate strategies and validated tools for implementation.

In an interesting parallel development, Singapore has been driving its 'Smart Nation Programme' that involves government agencies across various sectors from information communication and technologies to transportation and housing, to address the built environment through their focus on the development of big data and technology-driven innovations towards improving the future capabilities of transportation (Smart Mobility) and housing (Smart Living) (Hoe, 2016; Woo, 2017; Heng & Yeo, 2017). Under Smart Living, HDB has developed a 'Smart HDB Town Framework' that integrates smart technologies in the realm of environmental monitoring, estate management and building construction (Cheong, 2017).

Beyond physical infrastructure, it is possible that smart technology also helps to expand the socio-technical system (Ho, 2017). How might we build on the proliferation of smart applications among the citizens, leverage on data science and drive the current framework from 'Smart Nation' and 'Smart Town' down to the community level to introduce 'Smart Community' in the planning and design of living environment?

Coupling 'heartware' development with a data-driven approach hence leads to the 'Smart Community Design' proposed in this book. In the context of housing, we refer to *Smart Community Design as a developmental approach that uses smart technologies and data analytics to understand residents' diverse profiles and build on their social capital, to ensure equal access to opportunities and improve their quality of life.* Furthermore, the approach incorporates a data-driven, inclusive design process that aims to enable residents to co-create and co-manage common places and community facilities, supported by local authorities. Ultimately, the goals of Smart Community Design are to achieve stronger social integration, place identity, a sense of ownership and community resilience among the residents.

## Overview of the Book

This book is the result of four years of research from 2017 to 2021 and another two years of application, documentation and reflection. Under the original title 'New Urban Kampung', it was initiated as an interdisciplinary research programme that coupled data analytics with social behavioural studies to derive better insights and thereby more targeted interventions, as well as to develop a comprehensive set of participatory toolkits and digital platforms to help planners generate evidence-based neighbourhood designs, in order to foster inclusive and resilient communities. As mentioned earlier, three towns were selected for studies, namely, Toa Payoh, Jurong East and Punggol, each representing a different town planning ideology and at different phases of development, so that the findings would be potentially scalable to other towns or cities. The programme resulted in four interlinked outcomes: (1) redefined resident segmentation; (2) neighbourhood-based QoL and Quality of Place framework; (3) new tools for socio-environmental data-driven planning; (4) Community Enablement Playbook and prototypes (Figure I.3). They are presented in multiple chapters in this book, which is broadly categorised into three parts that aim to illustrate a clear process of data-driven design.

### Part I: Enquiring Diversity

The first part of the book outlines the formulation of two studies that lay the foundation for this research. Through multi-modal data collection methods and analytics, they provide the know-how to assess and segment the profiles of shifting residents' demographics and uncover their diverse perspectives, priorities and perceptions of QoL.

Beyond the conventional single-demographic analysis, Chapter 1 showcases the ability to redefine resident segments based on a multi-variable approach, i.e. simultaneously considers 14 socio-economic and lifestyle variables in two-level segmentation

**FIGURE I.3** Four interlinked outcomes of the New Urban Kampung (NUK) Research Program.

*Source:* Author.

analysis, to achieve a more complex profile of the population segments. Drawing data from more than 5,000 surveys conducted in three residential towns, 8 emerging resident segments and 20 sub-segments have emerged as a case study in Singapore, which offer a detailed understanding of their diverse perceptions and needs.

Chapter 2 presents a new QoL and Quality of Place framework and analysis at the neighbourhood level focusing on everyday lives of ordinary residents. This is contrary to most QoL studies done at the city level mainly for cross-city comparison. Based on surveys and 241 additional in-depth interviews with residents, the bottom-up data analytic process reveals the various factors that affect the residents' perception of QoL as well as Quality of Place, and what matters to them most, which can be measured through 67 objective and subjective indicators. The framework then allows planners to perform analytics comparing different towns, neighbourhoods and resident segments, and to derive targeted approaches to support residents' complex needs, which are explained in the following chapters.

## Part II: Envisioning Opportunities

By consolidating multi-modal data and uncovering the interdependencies between urban systems and social factors, four urban analytics tools and processes have been developed that can aid socio-environmental data-driven planning and help to identify intervention sites, targeted residents and areas of opportunities to better residents' QoL.

Chapter 3 illustrates the workflow of data analytics that are focusing on correlating community dimensions and planning parameters, such that it allows planners to examine the impact of precinct typologies and planning parameters on residents' perceived QoL.

Chapters 4 and 5 introduce various methods to analyse neighbourhood activities and residents' mobility. Chapter 4 focuses on utilising GPS and WiFi data fusion and clustering to identify participating residents' activities at diverse common spaces; while Chapter 5 outlines the methods of analysing passively sensed smartphone signals to visualise the mobility patterns of large groups of residents within the neighbourhood.

Social network analysis presented in Chapter 6 is a data-light model that could efficiently identify places where there is high potential for social interactions that is useful for site selection and place-making. Conversely, it can be used to identify potential crowded spaces to prevent overcrowding, which was especially useful during a health pandemic.

In Chapter 7, the development of a virtual population model provides an agent-based predictive analysis to simulate dynamically how the town population, resident segments and household composition will evolve with time. The model can be used to anticipate emerging needs and critical turning points.

These urban analytics tools and processes had been incorporated in a visualisation platform, City Application Visualisation Interface (CAVI) by EDF Singapore Lab, that enables planners and designers to make comparisons between towns, neighbourhoods and resident segments, to analyse proximity, accessibility and centrality of amenities and spaces, and to simulate emerging needs and opportunities for future community design.

### Part III: Enabling Inclusivity

Building on the deep understanding of diverse residential demography in Part I and the comprehensive urban analytics tools and processes in Part II, community-based participatory research was carried out in targeted intervention sites, which led to the development of a Community Enablement Playbook.

Chapter 8 elaborates on the Community Enablement Playbook which offers a suite of co-learning, co-creation and co-management strategies depending on the target resident segments and the spectrum of needs and assets in different neighbourhoods. With a data-driven capability, the playbook departs from conventional development through employing urban analytics processes to identify gaps, opportunities and suitable sites and programmes. The playbook draws lessons from case studies of community initiatives across Singapore as well as real-life engagement and implementations in the three study sites and showcases two successful place-making projects in collaboration with resident champions.

Despite the latter part of the research period being hit by the COVID-19 pandemic, the research team quickly rose to the challenge and seized the opportunity to innovate. While physical contacts were restricted, various digital tools were harnessed to develop new processes for community engagement. Chapter 9 begins with case studies of the digital landscape of community initiatives that surfaced during the pandemic, which led to the development of a one-stop online community platform that enables local stakeholders and resident champions, leveraging on offline-to-online/online-to-offline (O2O) engagement strategies.

Powered by new knowledge and tools, the concluding Chapter 10 reflects on the research programme, and discusses how the future of neighbourhood design, redevelopment and community living can be enhanced by more diverse offerings. Discussion is also extended to policy implications, which are illustrated through new housing policies targeted at an ageing demography, diversifying housing options, redefining town classification and community enablement. The chapter showcases selected housing studio projects from Singapore and Finland to highlight transferable knowledge and strategies, and to serve as inspirations to leverage on new data-driven strategies to create more inclusive and resilient neighbourhoods and communities.

### Note

1 The ROH plans for Punggol, despite being a young estate, included the development of a new Town Centre and a new 4.2-km man-made waterway meandering through the

town, further enhancing its identity as a waterfront town and offering more water-based activities. ROH would also bring new life into the middle-age Jurong especially the Jurong Lake District and Jurong East Town Centre, which would be rejuvenated into a new leisure and lifestyle destination and vibrant business hub with food and beverage, commercial, office, residential and hotel developments for travellers and residents. Toa Payoh was selected for a major make-over to inject new housing developments into the mature estate, in a later ROH programme that adopted a more ground-up approach through public engagement to solicit feedback and ideas.

## Bibliography

Channel News Asia (CNA). (2021, February 21). Ghim Moh resident sets up community children's library at her lift landing to share the joy of reading. *Channel News Asia (CNA)*. Retrieved from https://www.channelnewsasia.com/singapore/ghim-moh-resident-community-childrens-library-367696

Cheong, K. H. (2017). The evolution of HDB towns. In C. K. Heng (Ed.), *50 Years of urban planning in Singapore* (pp. 101–125). Singapore: World Scientific.

Chong, K. H., & Tan, D. (2015). *Designing communities: Evolutions of precinct public space design in Singapore public housing estates*. Gwanju: East Asian Architectural Culture.

Chua, V., Koh, G., Tan, E. S., & Shih, D. (2020). *Social capital in Singapore: The power of network diversity* (1st ed.). Oxon, New York: Routledge.

Cho, I. S., & Križnik, B. (2017). *Community-based urban development: Evolving urban paradigms in Singapore and Seoul*. Singapore: Springer.

Department of Statistics Singapore (2020). *Census of population 2020*. Retrieved from https://www.singstat.gov.sg/-/media/files/publications/cop2020/sr1/cop2020sr1.ashx

Fernandez, W., & Singapore Housing & Development Board (2011). *Our homes: 50 Years of housing a nation*. Singapore: Straits Times Press.

Goh, H. C., & Heng, C. K. (2017). Shaping Singapore's cityscape through urban design. In C. K. Heng (Ed.), *50 Years of urban planning in Singapore* (pp. 211–234). Singapore: World Scientific.

Heng, C. K. (Ed.) (2017). *50 Years of urban planning in Singapore*. Singapore: World Scientific.

Heng, C. K., & Chong, K. H. (2006). *Urban revitalization through housing induced public space*. Singapore: 4th Great Asian Street Symposium (GASS).

Heng, C. K., & Yeo, S. J. (2017). Towards greater sustainability and livability in an urban age. In C. K. Heng (Ed.), *50 Years of urban planning in Singapore* (pp. 287–303). Singapore: World Scientific.

Ho, E. (2017). Smart subjects for a smart nation? Governing (smart) mentalities in Singapore. *Urban Studies, 54*(13), 3101–3118.

Hoe, S. L. (2016). Defining a smart nation: The case of Singapore. *Journal of Information, Communication and Ethics in Society, 14*(4), 323–333.

Housing & Development Board (HDB). (1995). *Public housing design handbook*. Singapore: HDB.

Housing & Development Board (HDB). (2005). *Public housing design handbook*. Singapore: HDB.

Housing & Development Board (HDB). (2018). *HDB sample household survey 2018: Public housing in Singapore: Residents' profile, housing satisfaction and preferences*. Housing & Development Board (HDB). Retrieved from https://www.hdb.gov.sg/cs/

infoweb/-/media/HDBContent/Images/CDG/Library/Library/SHS-2018-Monograph-1—2-Mar-2021_BLUE.pdf

Housing & Development Board (HDB). (2020). *Home, truly: Building dreams, housing hopes.* Singapore: World Scientific.

Ooi, G. L., & Tan, T. T. W. (1992). The social significance of public spaces in public housing estates. In B. H. Chua & N. Edwards (Eds.), *Public spaces: Design, use and management* (pp. 69–81). Singapore: Centre for Advanced Studies/Singapore University Press.

Sanoff, H. (2006). Origins of community design. *Progressive Planning, the Magazine of Planners Network, 166* (Winter), 14–16.

Tan, E. S. (2017). Public housing and community development: Planning for urban diversity in a city-state. In C. K. Heng (Ed.), *50 Years of urban planning in Singapore* (pp. 257–272). Singapore: World Scientific.

The Straits Times (2022, March 19). Take a book, leave a book: Little community libraries spring up in S'pore. *The Straits Times.* Retrieved from https://www.straitstimes.com/life/arts/take-a-book-leave-a-book-little-community-libraries-spring-up-in-spore

The Straits Times (2023, January 30). Want to read books and make friends? Head for libraries set up by residents in HDB void decks. *The Straits Times.* Retrieved from https://www.straitstimes.com/singapore/want-to-read-books-and-make-friends-head-to-libraries-set-up-by-residents-in-hdb-void-decks

The Straits Times (2023, April 23). Boon Lay open library books stolen, then returned a day after shelves were taken. *The Straits Times.* Retrieved from https://www.straitstimes.com/singapore/boon-lay-open-library-books-stolen-then-returned-a-day-after-shelves-were-taken

The Straits Times (2023, October 20). Boon Lay void deck library to close due to complaints, messiness, lack of community effort: Founder. *The Straits Times.* Retrieved from https://www.straitstimes.com/singapore/boon-lay-void-deck-community-library-to-close-due-to-complaints-messiness-and-lack-of-community-effort-founder

The Straits Times (2023, November 6). I'd rather people take the books and not return them: Founder of community library in Tampines. *The Straits Times.* Retrieved from https://www.straitstimes.com/singapore/i-d-rather-people-take-the-books-and-not-return-them-founder-of-community-library-in-tampines

The Straits Times (2024, April 12). Singapore is world's 5th smartest city, top in Asia: Global index. *The Straits Times.* Retrieved from https://www.straitstimes.com/singapore/singapore-is-5th-smartest-city-in-the-world-top-in-asia-global-index

Urban Redevelopment Authority (URA). (2024) *Jurong gateway.* Retrieved from https://www.ura.gov.sg/Corporate/Guidelines/Urban-Design/Jurong-Gateway

Woo, J. (2017). *Singapore's smart nation initiative–A policy and organisational perspective.* Singapore: Lee Kuan Yew School of Public Policies.

Yeoh, B. S. A., & Kong, L. (1995). Portraits of places: History, community, and identity in Singapore. Singapore: Times Editions.

# PART I

# Enquiring Diversity

The first part of the book outlines the formulation of two studies that lay the foundation for this research. Through multi-modal data collection methods and analytics, they provide the know-how to assess and segment the profiles of shifting residents' demographics and uncover their diverse perspectives, priorities and perceptions of quality of life.

DOI: 10.4324/9781003437659-2

# 1

# INNOVATIVE DEMOGRAPHIC SEGMENTATION IN SINGAPORE

## A Multi-variable Approach for Policy and Community Planning

*W. Quin Yow and Shuang Zheng*

### The Imperative for a New Approach to Resident Demographic Segmentation in Singapore

In the context of Singapore's evolving demographic landscape, characterised by diversifying communities and a swiftly ageing populace, it becomes imperative to delineate population segment profiles and community characteristics that transcend traditional, broad socio-economic and demographic categorisations. This necessitates the formulation of bespoke, nuanced indicators. Such segmentation is crucial for the formulation of nuanced policy and planning strategies, ensuring they are meticulously tailored to meet the unique requirements of demographically varied communities.

Traditionally, we only look at single-demographic variables when analysing Singapore or for policy creation. An obvious variable often investigated would be 'age'. For example, it is commonly understood that older adults (i.e. individuals aged 65 and over) have different wants and needs, and as such would require specific policies to cater to those needs, such as clinics and accessible walking infrastructures. The goal of the resident segmentation in this work is to simultaneously consider multiple variables in order to gain a deeper understanding of differing wants and needs. For example, an older adult living alone with low income might need different government support compared to an older adult living with their children and grandchildren. To that end, a set of variables that included standard, more conventional socio-economic variables as well as behavioural and perception-based variables was considered for the segmentation analysis. Thus, this study aims to adopt a more holistic approach by examining multiple variables concurrently to capture the intricacies of diverse needs and preferences.

DOI: 10.4324/9781003437659-3

## Data-Driven Methodologies

### Developing a Contextual Model for Resident Segmentation in Singapore

The techniques and foundational principles of market segmentation are transferable to public service contexts. Existing segmentation frameworks typically integrate four types of variables: (1) geographic location; (2) demographics; (3) psychographics such as interests and values; and (4) behaviour patterns like purchase habits (Töpfer & Bug, 2015).

The research adopted a data-driven approach to variable selection, starting with a broad set of variables and narrowing down based on collected data. This process involved surveys and interviews to determine the most significant variables for segmenting Housing & Development Board (HDB) residents in Singapore. Table 1.1 in the original study provides an overview of existing models and their relevant variables, serving as a reference for the current project.

**TABLE 1.1** Segmentation variables in existing models and the corresponding variables used in the current research

| Segmentation Base Variable | Existing Segmentation Models Using These Factors | Relevant Variables Adopted in New Urban Kampung (NUK) Survey Questions |
| --- | --- | --- |
| Geographical location | • Acorn by CACI (UK)<br>• Claritas PRIZM Premier (USA)<br>• Mosaic by Experian (UK) | • Location of residence (Town, Neighbourhood, Precinct)<br>• Residence and geographical proximity to other amenities |
| Demographic information | • Tapestry Segmentation by ESRI (USA)<br>• Output Area Classification (OAC) by Office for National Statistics (UK) | • Basic demographics such as age, sex, income level, occupation, etc.<br>• Singapore-specific variables such as race, religion |
| Psychographic profiles | • Roy Morgan Values Segments (Australia)<br>• VALS (USA) | • Leisure activities (and where they do them)<br>• Personality<br>• Attitudes towards housing and satisfaction regarding HDB, etc. |
| Behavioural attributes | | • Participation in community activities<br>• Usage of residential facilities<br>• Awareness and knowledge of community events etc. |

*Source*: Authors.

## *Refining Segments of Residents through a Large-Scale Survey*

The first step involved brainstorming variables that were expected to impact geo-sociographic segmentation, Quality of Life (QoL), socio-spatial behaviours, as well as community participation[1] from a bottom-up and resident-centred perspective. Collaboratively, the research team created a long list of variables that were potential candidates to be included in the survey questions (Table 1.2).

**TABLE 1.2** Preliminary list of variables considered as potential candidates to be included into the NUK survey

| Category | Variables |
| --- | --- |
| Basic Socioeconomic status (SES) information | Address, Age, Gender, Household size, Size of Housing, Position in Household, Marital Status, Ethnicity, Language Use (i.e. dominant language, bilingualism proportion, self-rated proficiency), Educational Attainment, Employment/Occupation (sector/field, title/level, years of working experience or retirement), Personal Income |
| Personal mobility/ behaviour | Hours Worked per Week, Location of Employment, Travel mode/ time, Hours Spent at Home (excluding sleep), Consumer Behaviour (i.e. recreational activities/hobbies, hours spent on them per week, monthly household expenditure, common shopping places, common dining places), Community Participation (i.e., participation in community activities, use of common facilities) |
| Household-level information | Type of Household, Household Particulars (i.e. age, gender, position in household, ethnicity, educational attainment, employment/ occupation), Monthly Household Expenditure, Length of Stay |
| Community & infrastructure | Satisfaction levels and possible improvements in existing areas such as Common Corridors, Void Decks, Community Spaces (i.e. garden, multi-purpose courts/halls), Wheelchair/Pram Accessibility, Noise Levels, Weather Conditions (i.e. heat/ rain-related infrastructure such as covered walkways, solar panels), Parking, Cycling Paths, General Safety, Accessibility to Healthcare (i.e. distance to clinics), Environmentally Sustainable Practices (i.e. recycling bins, programmes), Amount of Landscapes (i.e. greenery) |
| Factors Affecting Housing Choice | Location, Accessibility, Nearby Amenities/Buildings (including both malls and religious places), Weather Concerns (i.e. direction relative to wind, direction relative to sun), Floor plan Layout, Affordability, Noise, Size of House, Environmental Sustainability |
| Individual Psychological Variables | Personality (Big 5), Satisfaction with Life (General SWLS – domain specific in areas such as health, financial situation, social relationships, self-worth, leisure time, family, work, education, healthcare, housing, public utilities, transport, environment, religion, politics), Sense of Neighbourliness, Perceived Ownership of Common Facilities, Openness to Community Sharing/ Cooperation, Perceived Friendliness of Neighbours |

*Source*: Authors.

Using the list of potential variables, survey questions were generated or adopted from well-established scales. These sections of the survey were based on the following scale:

- Section A10: Language background (adapted from Yow & Li, 2015)
- Section A12: Mother tongue (MT) language perception (adopted from Surrain & Luk, 2019)
- Section C2: General life satisfaction (extracted from Satisfaction with Life Scale; Diener et al., 1985)
- Section C5: Attitudes towards elderly (adapted from Kogan's attitudes towards older people scale, with reference to Yen et al., 2009)
- Section C7: Perception of technology (adapted with reference to Chen & Chan, 2014)
- Section G5, H2a, H3, H6, and H7: Sense of belonging to neighbourhood (selected items adapted from Woodcraft & Dixon, 2013)

Table 1.3 shows the groupings of variables and sub-sections of questions under each section. Variables that were open-ended or required clarification and deemed as better explored using other data collection methods were also removed from the survey and included in the pilot interviews instead.

**TABLE 1.3** Overview of survey sections

| Section | Sub-section(s) |
| --- | --- |
| A: Demographic information | • Type of housing<br>• Basic demographic information (e.g. age, sex, ethnicity, religious inclination)<br>• Educational attainment<br>• Language background<br>• MT language perception |
| B: Occupation/ work life | • Employment status & details (e.g. type, sector, monthly household income)<br>• Work & work-life balance<br>• Personal development |
| C: Individual preferences | • Personality<br>• General life satisfaction<br>• Perception of health<br>• Attitudes towards elderly<br>• Perception of social changes<br>• Perception of technology |
| D: Personal lifestyle & habits | • Leisure (e.g. locations, activities, frequency, social circle)<br>• Perceptions of social life |

(*Continued*)

**TABLE 1.3** (Continued)

| Section | Sub-section(s) |
|---|---|
| E: Consumption patterns | • Satisfaction towards household expenditure |
| F: Housing and family information | • Tenure status<br>• Family ties and relationships<br>• Perceived comfort<br>• Housing satisfaction<br>• Perception of HDB and housing |
| G: Neighbourhood perceptions | • Preferred mode of transport<br>• Satisfaction towards accessibility<br>• Neighbourhood facilities satisfaction<br>• Sense of belonging towards neighbourhood |
| H: Community interactions | • Perception of neighbours<br>• Neighbourly interactions<br>• Neighbourly relations<br>• Community contribution<br>• Governance and perception of influence |

*Source*: Authors.

This study undertook adjustments to the scale structure from previous scales and simplified them based on factor loadings. Subsequently, feedback was solicited from research collaborators including the HDB and EDF Lab Singapore to refine the questionnaire. For instance, the term 'shabby' was considered overly negative and inappropriate for the survey context. The question was reversed into the statement, 'The elderly have clean, attractive homes' as an alternative.

The survey was further tailored with new items designed from prior knowledge and pilot interviews. Results from preliminary interviews informed the exclusion of uniform responses and the update of key QoL items reflecting domains frequently highlighted by participants.

## Survey Results and Analysis

### Community Survey

Prior to data collection, ethics approval was obtained from SUTD-IRB on September 5, 2018. Then, two forms of community surveys were conducted: a comprehensive door-to-door survey and a shorter street intercept survey (Table 1.4). The former aimed for an in-depth understanding of 3,000 residents, requiring 45 minutes, while the latter sought a broader understanding of public opinion regarding the location, the available amenities, and the perception of that residential town from 2,000 individuals across sites, requiring only 5 minutes. People who were non-residents of the town and/or non-HDB residents were allowed to participate. Prior to full-scale data collection, a two-week pilot survey was undertaken in May 2019, leading to minor revisions based on feedback from 90 respondents across the towns.

**TABLE 1.4** Survey sites for street intercept survey

| No. | Town | Location |
| --- | --- | --- |
| 1 | Toa Payoh | HDB Hub |
| 2 | | Toa Payoh West Market & Food Centre |
| 3 | Punggol | Waterway Point |
| 4 | | Oasis Terraces |
| 5 | Jurong East | J-Link |
| 6 | | Yuhua Market & Hawker Centre |

*Source*: Authors.

## Sampling Strategy

A systematic selection of households, comprising a stratified sample of 7,500 units, was used for the door-to-door survey. Sampling was randomised using the last-birthday method to include different household members, not necessarily owners. For the street intercept survey, no quotas were set, and every fifth passerby was approached at selected high- and low-traffic sites at varied times.

## Data Collection and Methodology

Between July and November 2019, 7,500 households were invited by mail to a voluntary survey, offering $40 vouchers as an incentive. Households had to be occupied for a minimum of three months, with inclusion regardless of ethnicity or nationality.

## Survey Execution

Surveys, lasting 45 minutes, required signed consent forms. Interviewers recorded responses on tablets and paper, offering translations. It featured multiple-choice and open-ended questions. Incentives were distributed post-survey with receipt confirmation, and consent for future contact was obtained when permitted.

## Data Security and Quality

Data was stored on a secure online platform, encrypted throughout the collection and transmission processes. Additional safeguards included password-protected tablets, restricted data access, and ongoing quality checks for data integrity and interviewer performance.

## Survey Findings Overview

The survey amassed 3,134 responses distributed across Jurong East, Punggol, and Toa Payoh. Checks confirmed sample representativeness in terms of household types, geographic distribution, and demographics. The sample mainly comprised Singaporeans. Income levels and educational attainment varied across regions. Overall, the sample was deemed to align well with the general population demographics of the surveyed towns (Figures 1.1–1.4).

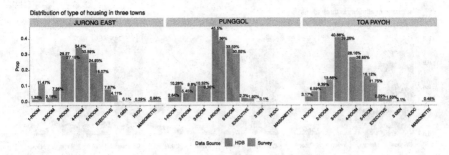

**FIGURE 1.1**   Type of housing distribution comparing HDB data and survey sample.

*Source*: Authors.

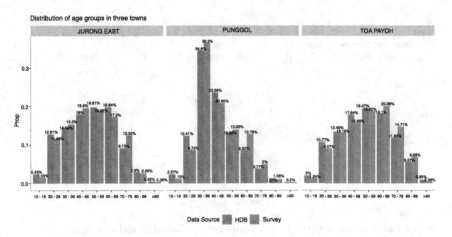

**FIGURE 1.2**   Age group distribution comparing HDB data and survey sample.

*Source*: Authors.

### *Street Intercept Survey*

The street intercept survey (Table 1.5), executed in August and November 2019, aimed for a representative sample of visitors at various sites. Multiple interviewers worked in shifts, targeting 42 respondents per session from people aged 18 and up, excluding certain groups like tourists and shopkeepers. The survey, which took approximately 5 minutes to complete, assessed the participants' reasons for visiting and their impression of the locale.

In total, 2,021 participants completed the survey, distributed across three different towns: Toa Payoh (672), Punggol (677), and Jurong East (672). Age distribution corresponded to the general demographics of the towns. Most respondents frequented these sites due to proximity and convenience.

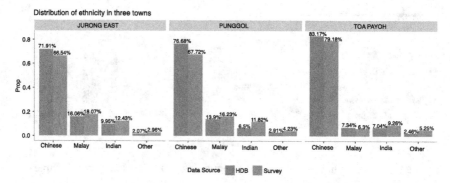

**FIGURE 1.3** Ethnicity distribution comparing between HDB data and survey sample.
*Source*: Authors.

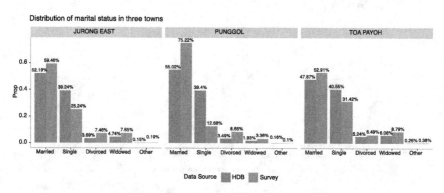

**FIGURE 1.4** Marital status distribution comparing HDB data and survey sample.
*Source*: Authors.

**TABLE 1.5** Breakdown of final survey sample for street intercept survey

| No. | Town | Location | Respondents |
| --- | --- | --- | --- |
| 1 | | HDB Hub | 336 |
| 2 | Toa Payoh | Toa Payoh West Market & Food Centre | 336 |
| 3 | | Waterway Point | 340 |
| 4 | Punggol | Oasis Terraces | 337 |
| 5 | | J-Link | 336 |
| 6 | Jurong East | Yuhua Market & Hawker Centre | 336 |

*Source*: Authors.

## Segmentation and Model Verification

### *Demographic Classification*

The variables were shortlisted based on importance. Prior to the segmentation analysis, the data was cleaned and processed, resulting in a total of 28 variables comprising a mixture of demographic/socio-economic and perception/behaviour-based variables that would be the basis of the final attribute selection in the segmentation analysis (Table 1.6).

**TABLE 1.6** List of variables to be used for segmentation analysis

| ID | Variable Type | Variable Name |
|----|---------------|---------------|
| 1  | Demographic   | Age |
| 2  |               | Ethnicity |
| 3  |               | Highest education attained |
| 4  |               | Religion |
| 5  |               | Employment status |
| 6  |               | Employment field |
| 7  |               | Job time |
| 8  |               | Household income per month |
| 9  |               | Times you have moved house over lifetime |
| 10 |               | Number of years lived in the house |
| 11 |               | Tenure status |
| 12 |               | Household type |
| 13 |               | Number of people living in the house |
| 14 |               | Average percentage of time using language |
| 15 | Perception/   | Average perception towards work and income |
| 16 | behaviour     | Average perception towards work-life balance |
| 17 |               | Average perception towards personal development |
| 18 |               | Average perception of health |
| 19 |               | Average perception of social changes |
| 20 |               | Average attitudes towards elderly (appreciation and prejudice) |
| 21 |               | Average perception of technology (usefulness, anxiety and self-efficacy) |
| 22 |               | Average frequency for visiting places for leisure (Shopping and F&B; Sports & Play; Social & Community facilities) |
| 23 |               | Do you generally go to these places within or outside your neighbourhood? |
| 24 |               | Do you intend to move out in the next five years? |
| 25 |               | Family members/Relatives living within ten minutes' walk |
| 26 |               | Average satisfaction with accessibility to bus stops and Mass rapid transit (MRT) stations |
| 27 |               | Average perception of neighbourly interactions |
| 28 |               | Average perception of community contribution |

*Source*: Authors.

### K-Medoids

K-medoids, also known as PAM (Bhat, 2014), is preferred for mixed data type segmentation due to its robustness, surpassing k-means in handling categories like household composition. It identifies k representative objects (medoids) that minimise intra-cluster dissimilarity, measured via Gower's distance (Gower, 1971), effectively grouping similar observations, and distinguishing them from other clusters.

### Variable Selection

Prior to clustering, variables underwent cross-correlation to mitigate multicollinearity. After the discussion, it was decided that a two-level hierarchical clustering would be a better approach. The first tier utilised a condensed set of influential variables conducive to clear segmentation and potential replication across broader populations. The second tier involved further subdivision using additional variables. Notably, 'ethnicity' was excluded for its non-convergence in clustering, while 'language spoken', quantifiable by usage percentage, was preferred for analytical coherence (Table 1.7).

### Identifying the Optimal Number of Clusters

After data cleaning, the study's subject pool shrank from 3,134 to 2,580. Utilising k-medoids clustering with an optimal $k = 8$, the study achieved a high average silhouette width, denoting well-defined clusters (Figure 1.5).

We visualised the eight resident segments on a two-dimensional space (Figure 1.6), where the distance between respondents indicated the extent of dissimilarities between the two. Respondents were coloured based on the segments they were assigned to in order to provide a visualisation of segments that grouped together in the same 'corner' of the graph.

**TABLE 1.7** List of variables for first-level clustering

| ID | Variable Name |
| --- | --- |
| 1 | Age |
| 2 | Highest education attained |
| 3 | Employment status |
| 4 | Household income per month |
| 5 | Number of years lived in the house |
| 6 | Tenure status |
| 7 | Household type |
| 8 | Percentage of time using language |

Source: Authors.

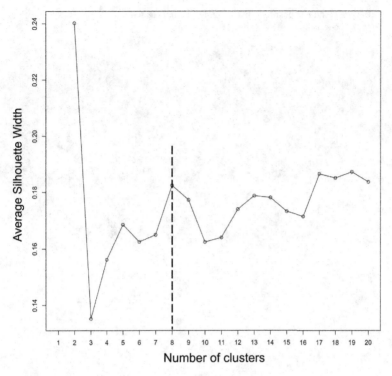

**FIGURE 1.5**   Average silhouette width plot for determining optimal number of clusters in the first-level clustering.

## First-Level Clustering

### Results of the First-Level Clustering

Table 1.8 and Figure 1.7 outline and summarise the eight resident segments derived from the first-level clustering. Each segment was issued a name followed by detailed descriptions based on the statistical summary of selected variables.

Figure 1.8 illustrates the spatial distribution of eight resident segments across three towns, revealing a higher presence of 'Golden Seniors' and 'Silver Contributors' in Toa Payoh, affirming its status as a mature estate. In contrast, 'Modest Tenants' and 'Independent Lessees+' are predominantly located in Jurong East N3a and Punggol Field East, respectively. Diverse resident characteristics across neighbourhoods are inferred from these distributions.

Figure 1.9 shows the demographic and socio-economic characteristics across the eight resident archetypes. It covers a range of variables as what we investigated including household composition, age distribution, income levels, educational attainment, employment status, housing tenure, and engagement with multiple languages, reflecting Singapore's multicultural society.

**FIGURE 1.6** Cluster analysis of survey respondents using PAM with Gower distance in a reduced dimensional space.

*Source*: Authors.

Figure 1.10 shows that the most common job type in the dataset is 'Professional', followed by 'Service, shop or market sales worker'. In contrast, 'Public Service Worker' and 'Craftsman and related trades worker' have the smallest proportions. In the bottom chart, we observe that the distribution of job types varies by town:

- 'Professional' jobs make up a larger proportion in Toa Payoh compared to Jurong East and Punggol.
- 'Service, shop, or market sales worker' jobs are relatively evenly distributed across the three towns, with a slightly higher proportion in Jurong East.
- Jurong East has a noticeably larger proportion of 'Associate Professional/ Technician' jobs compared to the other two towns.

**TABLE 1.8** Descriptions of resident segments derived from first-level clustering

| ID | Name | Percentage | Highlight | Description |
|---|---|---|---|---|
| C1 | Golden Seniors | 13.9% (358) | Retirees with low income | This cluster consists of an elderly population (avg. age 71) that is retired (78%). Many primarily speak Chinese (35%) or one of the dialects (43%). They have lived in the same house, which they primarily own (83%), for a long time (avg. 24 years). They have relatively low incomes (68% are below $1,000) and educational levels (64% are primary and lower). |
| C2 | Secure Homesteader | 12.3% (318) | Middle-aged adults with adult children | This is a cluster of middle-aged people (avg. age 56) with adult children (78%). Many primarily speak Chinese (30%) or one of the dialects (21%). They are employed (72%) and generally have not received a university education (48% are secondary). Most have lower to middle income. They own their own home, in which they have generally lived a long time (avg. 17 years). |
| C3 | Modest Tenants | 9.5% (246) | HDB renters | Although the average age of this cluster is 46, the standard deviation is very high (14 years). This cluster has a high percentage of Malay speakers (36%). They are generally employed (63%), although a portion is unemployed. Incomes are very modest (76% are lower than $2,000), as are education levels (79% are secondary or primary and lower). They rent their home directly from HDB and only have lived in them for a small number of years (avg. seven years). |

*(Continued)*

**TABLE 1.8** (Continued)

| ID | Name | Percentage | Highlight | Description |
|---|---|---|---|---|
| C4 | Empowered MillennialZ | 8.8% (228) | Young adults yet to own HDB | This cluster is young (avg. age 33, sd 14 years). It has a large percentage of Chinese speakers (38%). Many are employed (64%) but a significant portion is (still) studying full-time. Most of the cluster consists of household types consisting of a couple with adult children (54%). The tenure status makes clear that the people actually in this cluster are the adult children ('House is owned: I am family relation of owner'). This cluster has a high degree of polytechnic and university degrees (73%). |
| C5 | Silver Contributors | 13.8% (355) | Older adults (many single or w/o children), about to retire | This cluster consists of an elderly population (avg. age 63). They are similar to Cluster 1 (also Chinese/dialect speaking) with an important difference, they are mostly not retired yet (65%). Perhaps as a result, their incomes are slightly higher than Cluster (46% are below $1,000). This group is potentially fragile in future years as they mostly consist of single-person households (79%). Most own their own house (82%) but a portion rent from HDB (15%). |
| C6 | CoFam+ | 18.3% (473) | Young parents with high incomes | This is the (upper) middle-class cluster. The average age is 39. Chinese (but not dialect) is often spoken (41%). Most are employed (87%) with some stay-at-home parents. They predominantly have university degrees (61%) and have high incomes (21% are $10,000 and above). Most are couples, with the majority also having children (<18 years old) (74%). They own their home but have spent relatively few years in their current home (avg. six years). |

*(Continued)*

**TABLE 1.8** (Continued)

| ID | Name | Percentage | Highlight | Description |
|---|---|---|---|---|
| C7 | Independent Lessees+ | 6.0% (156) | Young open-market renters | They are young (avg. 35), employed (83%), with high education (77% have university degrees) and income levels (20% are $10,000 and above). There are a few striking differences. This cluster is renting from someone else (not HDB), and they have only been in their home for three years. Although some are married couples with young children (28%), many are single-person households (41%). This is also the only cluster that has a relatively high percentage of Tamil speakers (9%) and a lower percentage of Chinese speakers. If we would map the spatial distribution of this cluster, many will likely be located in Punggol. |
| C8 | Multi-Gemmers | 17.3% (446) | Younger middle-aged adults in multi-generational households | This cluster is very similar to Cluster 6, with similarities to Cluster 4 as well. Generally, youngish (avg. age 45) with Chinese a commonly spoken language (38% of time). They are also employed (86%), well educated (31% have a university degree, but lower than Cluster 6) and high income (21% are $10,000 and above, but lower than Cluster 6). The key difference is that they overwhelmingly (88%) live in multi-generational households. They own their home (86%). In comparison with Cluster 6 they have lived in their house longer (avg. 11 years). |

*Source:* Authors.

**FIGURE 1.7** Summary of resident segments derived from first-level clustering.

*Source*: Authors.

Employment-related variables were found to be significant in the second-level clustering, which will be elaborated in the next section.

## Second-Level Clustering

### Challenges Using Perception and Behaviour-Based Variables

Subsequent k-medoid clustering within identified resident segments faced challenges due to the homogeneity in perceptions among older Singaporeans, unlike the varied younger counterparts, leading to indistinct sub-clusters among groups like Golden Seniors and Secure Homesteaders (Figure 1.11).

Another obstacle of this approach manifested in the inability to use uniformed perception variables to group the entire population. In order to tackle the problem, a feature selection was conducted to identify the variables that played the most important roles in distinguishing sub-clusters. Variables with standard deviations below 0.8 were excluded, leaving behind seven variables. These seven variables were joined by the employment-related variables (i.e. 'Job Time' and 'Job Field') as well as three more binary variables (Table 1.9).

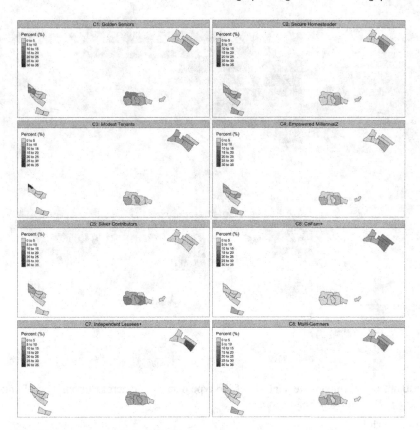

**FIGURE 1.8** Spatial distribution of first-level clusters across three towns.

*Source*: Authors.

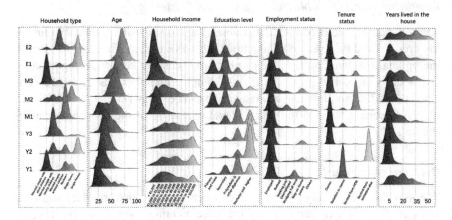

**FIGURE 1.9** Density distribution and statistical summary of individual variables within each cluster.

*Source*: Authors.

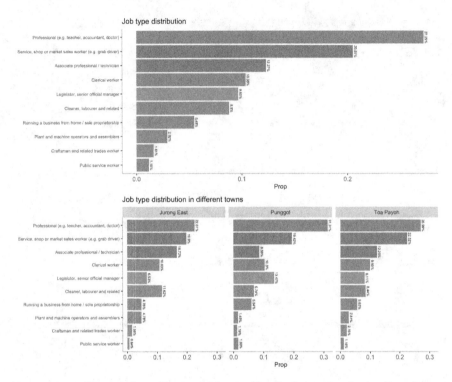

**FIGURE 1.10** Comparative analysis of job type distribution across Jurong East, Punggol, and Toa Payoh.

*Source*: Authors.

## Identifying Stable Variables for Second-Level Clustering

Random combinations of the 12 variables went through the k-medoid clustering process, and the average silhouette width was calculated in order to identify
the optimal combination (i.e. the combination with the largest average silhouette width value). The results from 4,083 unique combinations of the 12 variables found that employment-related variables (i.e. 'Job Time' and 'Job Field')
played important roles in the second-level clustering of the resident segments
(Table 1.10).

## Overview of Sub-clusters

The k-medoids clustering process was conducted for the eight resident segments
from the first-level clustering using the three additional variables. The average
silhouette width was calculated to determine the optimal number of sub-clusters
within each resident segment (Figure 1.12).

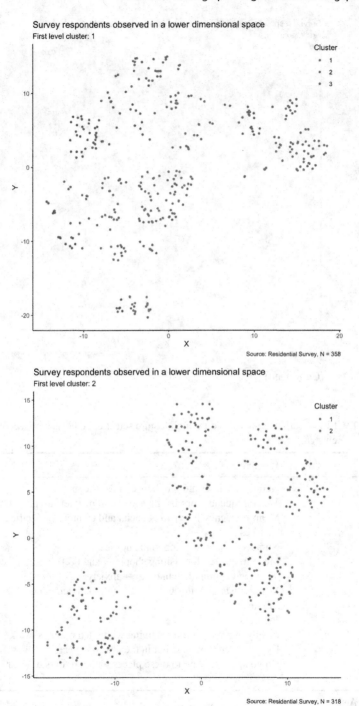

**FIGURE 1.11** Distribution of survey respondents on a lower dimensional space.

*Source*: Authors.

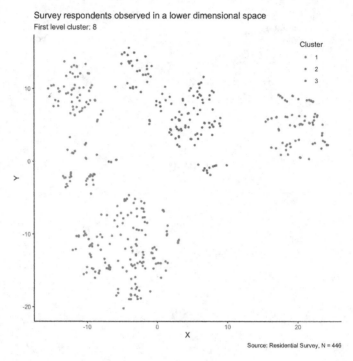

FIGURE 1.11 (Continued)

**TABLE 1.9** List of variables shortlisted through feature selection to be used for second-level clustering

| ID | Variable Name |
|---|---|
| 1 | Number of times moving house in the lifetime |
| 2 | Mean frequency for visiting sports and play leisure places |
| 3 | Mean frequency for visiting social and community facilities leisure places |
| 4 | Number of people living in the house |
| 5 | Mean frequency for visiting shopping and F&B leisure places |
| 6 | Mean perception of technology – anxiety |
| 7 | Mean satisfaction towards neighbourhood interaction |
| 8 | Job time |
| 9 | Job field |
| 10 | Family members/relatives living within ten minutes' walk |
| 11 | Do you intend to move out in the next five years? |
| 12 | Do you generally go to these places within or outside your neighbourhood? |

*Source*: Authors.

**TABLE 1.10**  Finalised list of variables for second-level clustering

| ID | Variable Name |
|----|---------------|
| 1 | Job time |
| 2 | Number of people living in the house |
| 3 | Family members/relatives living within ten minutes' walk |

*Source*: Authors.

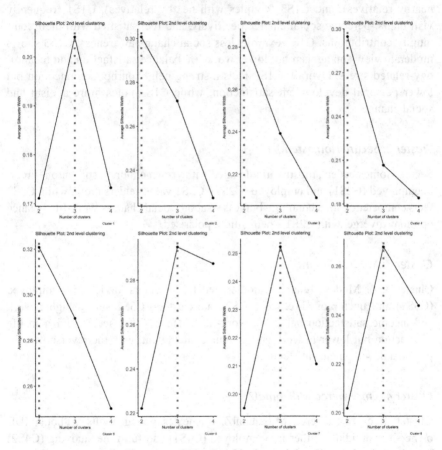

**FIGURE 1.12**    Average silhouette width plot for determining the optimal number of clusters in the second-level clustering.

*Source*: Authors.

## Comparisons Using Perception and Behaviour-Based Variables

### Comparing between Sub-clusters

The following comparisons were made with reference to the density distribution graphs of reported perception and behaviour-based variable scores across the sub-clusters of each resident segment and the diverging bar graphs comparing all sub-clusters across all resident segments.

### Cluster 1: Golden Seniors

Golden Seniors (Cluster 1), low-income retirees, are segmented into sub-clusters: C1S1 (couples without nearby relatives), C1S2 (individuals without nearby relatives), and C1S3 (couples with nearby relatives). C1S1 frequently visit malls, perceive social change positively, and feel satisfied with their community contributions. C1S2 exhibits less social change awareness. C1S3 shows moderate views on ageism but lower work-life balance satisfaction and technology-related stress. Overall, C1S2 feels a strong neighbourhood connection but lower personal development satisfaction, while C1S3 notes more ageism and social change.

### Cluster 2: Secure Homesteaders

Secure Homesteaders, mostly middle-aged with grown children, split into retirees/unemployed (C2S1) and employed (C2S2). C2S1 values elders more, while C2S2 feels competent with technology but less so about social change. Work-life balance opinions diverge, with C2S1's being more polarised.

### Cluster 3: Modest Tenants

Cluster 3, or Modest Tenants, middle-aged HDB renters, divide into employed (C3S1) and unemployed/retired (C3S2) sub-clusters. C3S1 shows higher work and income satisfaction but lower work-life balance. C3S2 scores higher in life satisfaction but lower in work and income contentment, with the reverse true for personal development satisfaction.

### Cluster 4: Empowered MillennialZ

Cluster 4, or Empowered MillennialZ, delineates young adults without HDB ownership, dividing further into employed (C4S1) and full-time studying (C4S2) sub-segments. C4S1 displayed varied satisfaction with work, income, and balance, but lower health perceptions and social satisfaction than C4S2, who scored average more consistently. C4S1 excelled in technological self-efficacy.

### Cluster 5: Silver Contributors

Cluster 5, or Silver Contributors, nearing retirement and often without children, comprise two subsets. The younger, employed sub-cluster, C5S1, shows less prejudice towards the elderly and more contentment with neighbours, frequents malls and eateries moderately, and expresses satisfaction with work and accessibility. C5S2, older and not working, reports better work-life balance. Across sub-clusters, C5S2 visits sports venues least but frequents community centres most.

## Cluster 6: CoFam+

Cluster 6, or CoFam+, delineates young, high-income parents, with sub-clusters C6S1 (regularly employed), C6S2 (stay-at-home), and C6S3 (irregular hours employed). Satisfaction metrics vary: C6S1 and C6S3 score average to high on job satisfaction and income, but lower on work-life balance. Contrastingly, C6S2 shows broad satisfaction distribution and highest respect for elders. Notably, C6S1 is least satisfied with neighbour interactions, C6S1 struggles with technology and accessibility, while C6S3 has minimal elder appreciation and senses little social change.

## Cluster 7: Independent Lessees+

Cluster 7, or Independent Lessees+, comprises young renters with three subgroups: employed regular workers (C7S1), unemployed stay-at-home parents (C7S2), and solitary irregular workers (C7S3). C7S1 frequents shopping and recreational venues moderately, while C7S2 shows low bias against elders and high personal development contentment. C7S3 scores low in neighbourly satisfaction but high in tech efficacy. Overall, C7S1 leads in community, work, and life satisfaction; C7S2 excels in neighbourhood interaction and health perceptions but lags in work-life balance; C7S3 leads in technology use but feels less community connectedness.

## Cluster 8: Multi-gemners

Cluster 8, comprising younger to middle-aged adults in multi-generational homes, splits into three: employed (C8S1), retired/unemployed (C8S2), and irregularly employed (C8S3) individuals. C8S1 showed lower perceptions of social change but higher work satisfaction, while C8S2 had moderate views on ageism and neighbour relations, with less work and income contentment. C8S2's work-life balance ratings varied widely, unlike the neutral tendencies of C8S1 and C8S3, who frequented shopping and dining venues more.

## Dimension Reduction for a Holistic Comparison

Employing Principal Component Analysis (PCA), a dimension reduction technique, this study distilled numerous perception and behaviour variables into two principal components explaining 58.3% of the dataset's variance. This analysis clustered variables by correlation, leading to two dimensions: the first encompassing Technology Perception and Work-Life Balance and the second combining Life

Satisfaction and Behaviour. This approach ensured a more comprehensive under-standing than individual variable comparisons could provide.

The study's 20 sub-clusters, represented as leaves, were organised into four groups based on similarity, as depicted in a dendrogram (Figure 1.13). In the group dendrogram, each sub-cluster is considered a single element, known as a leaf. Sub-clusters that are more similar to one another are combined into bigger groups, known as nodes. Group 1 (blue) mainly includes older adults but with one distinct sub-cluster, C1S2, indicating Golden Seniors living alone who may need special policy consideration. Group 4 (green) combines diverse sub-clusters like stay-at-home parents and both employed and unemployed Independent Lessees, unified by similar views on technology and work-life balance. Groups 2 (red) and

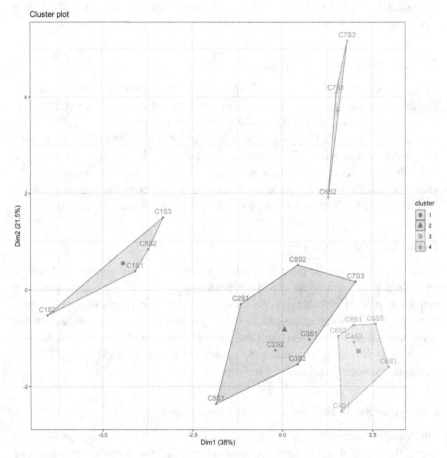

**FIGURE 1.13** Cluster plot depicting sub-clusters grouped based on perception and behaviour-based variables.

*Source*: Authors.

3 (yellow) encapsulate a broad mix of perceptions and behaviours, highlighting the distinct needs of older versus younger demographics for policy development. Moreover, sub-clusters such as C8S2 (retired Multi-Gemners) and C3S2 (employed Modest Tenants) show similar technology and work-life balance perceptions yet diverge in behaviour and life satisfaction, possibly due to the former's greater economic freedom.

By comparing perception and behaviour-based variables across sub-clusters of the resident segments, it can be observed that there are certain perceptions that several sub-clusters share. These resident segments are a way of simplifying the approach of understanding large groups of individuals by first dividing them up into meaningful groups (Figure 1.14). Hence, understanding the perceptions and behaviours of these resident segments is complementary to the process of planning effective and creative policies.

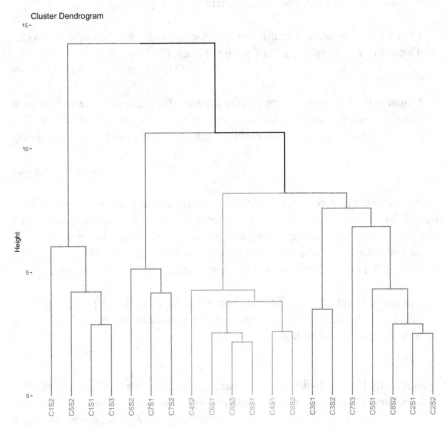

**FIGURE 1.14**  Cluster dendrogram depicting sub-clusters grouped based on perception and behaviour-based variables.

*Source*: Authors.

## Verification of the Model

We can verify the current model of resident segments and sub-clusters using excerpts from the interviews (refer to Chapter 2). Utilising the relatively robust demographic variables as the basis for the first-level and second-level segmentations, we can identify respondents who are typical examples of the derived resident segments and sub-clusters. These respondents' views can then be used to validate the current model.

## Comparisons Using Group 3 as an Example

Group 3, consisting of Sub-clusters C4S1, C4S2, C6S1, C6S3, C8S1, and C8S3, scored higher on Dim-1 (perception of technology & work-life balance) compared to other groups. Group 3 can be described as working adults and should include individuals. The following excerpts were extracted from respondents of the interviews that would have fallen within Group 3:

> I like to surf the net … I just like to visit social media, or read random news … This is my way of relaxing. Surf the net. Watch TV.
>
> *PH2TP03, 29, F, Assistant Manager*

> I think it is very important [to use the Internet] because next time will be even more technologically advanced and you have to be savvy with these things to continue. If not, you get stuck there [and then depend] on other people to help you [do it].
>
> *PH2TP32, 39, M, Private Hire Driver*

Many respondents from Group 3 addressed technology as a regular feature of their daily lives. Whether it is for leisure or for general day-to-day activities, it appeared that respondents from Group 3 generally perceived technology in a positive light, consistent with the patterns observed in the model. On the other hand, it appeared that Group 3's outlook on work-life balance differs from the observed trend:

> The work is very fast-paced and there are a lot of different projects that we manage. In terms of timeline, sometimes it's very tight. In terms of manpower, it is also usually not enough.
>
> *PH2TP03, 29, F, Assistant Manager*

> Not really [satisfied with work-life balance], because we have to work at night sometimes. There are conference calls at night time too.
>
> *PH2PG21, 37, F, Procurement Manager*

Respondents from Group 3 seemed dissatisfied with their work-life balance. While this contradicts the observation that Group 3 scored higher on Dim-1, it should be pointed out that Dim-1 was the result of dimension reduction in which highly correlated variables loaded onto the same principal component. Upon further comparisons solely focusing on work-life balance, it was found that the sub-clusters of Group 3 all reported below-average scores, consistent with the interview responses.

## Discussion and Conclusion

### Civic Engagement, Sense of Belonging among Various Segments

The imperative for a nuanced demographic segmentation in Singapore has never been more pressing, as evidenced by our current study. Our study builds upon this imperative, delving into the intricate dynamics of neighbourliness, civic engagement, and the sense of belonging among diverse population segments, with a particular focus on the role of MT as a critical factor. Our analysis of resident demographics illuminates the diverse needs within the population, prompting a re-evaluation of policy and planning frameworks.

Moreover, the study highlights the importance of considering the spatial distribution of demographic segments and their unique profiles in urban planning. As Singapore continues to develop its urban landscape, ensuring that the infrastructural and social needs of its diverse population are met is crucial. This includes designing spaces that encourage intergenerational interaction and cater to multilingual communities, thereby fostering a sense of belonging and civic engagement.

Figure 1.15 provides a comparative visualisation of community engagement and satisfaction across various resident archetypes. Archetypes with consistently green bars across charts could be considered as having strong community engagement and satisfaction, while those with red bars might identify areas needing attention or improvement in community initiatives and infrastructure.

From Figure 1.15 and the content above, the perception of neighbourliness in Singaporean society is multifaceted, as evidenced by the four distinct types identified by the research team[2]:

- Hi-Bye Neighbourliness
- Convenience Neighbouring
- Activity-Driven Neighbouring
- Community Contribution

Our findings suggest that these types manifest differently across various demographic segments, influenced significantly by the MT identity. English-speaking residents often display a higher propensity for 'Hi-Bye Neighbourliness', indicative of a more superficial form of interaction, while non-English MT speakers show stronger tendencies towards 'Activity-Driven Neighbouring' and 'Community Contribution', reflecting deeper communal ties.

Civic engagement is another domain where MT plays a consequential role. The study found that residents who communicate in their MT are more likely to participate in civic activities, suggesting that language can both empower and create barriers to civic participation. In this analysis, we compared different aspects of neighbourhood interactions and community engagement across various MT identities using univariate ANOVA. Specifically, we examined convenience neighbouring, activity-driven neighbouring, community contribution, and sense of belonging.

- *Convenience Neighbouring:* The analysis revealed significant differences in how different language groups engage in daily casual interactions with their neighbours, such as greetings and small talk. ($F(3, 713) = 7.22$, $p < 0.001$). For example, residents whose primary language is English tend to have more superficial interactions compared to those who speak other MTs.
- *Activity-Driven Neighbouring:* The results indicated that participation in community events and shared activities varies notably among language groups

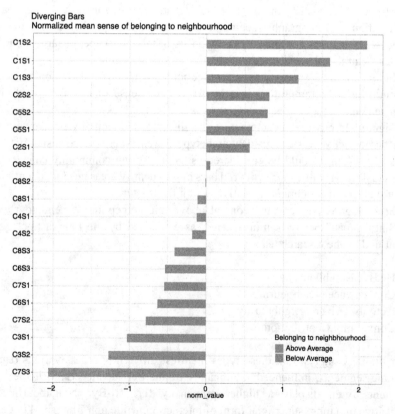

**FIGURE 1.15**  Comparative analysis of community engagement and satisfaction across resident archetypes.

*Source*: Authors.

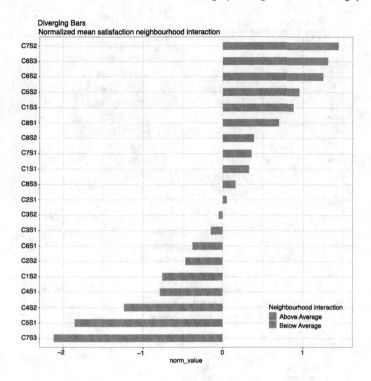

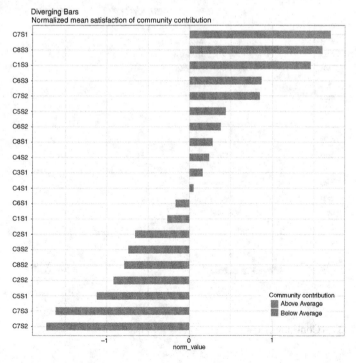

**FIGURE 1.15** (Continued)

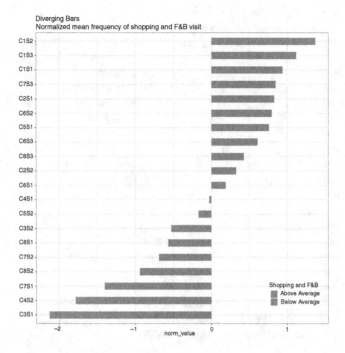

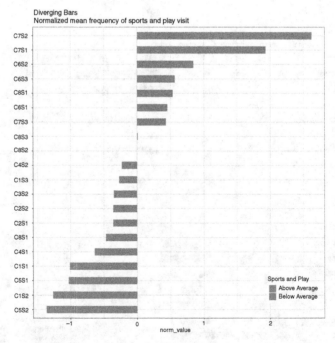

**FIGURE 1.15** (Continued)

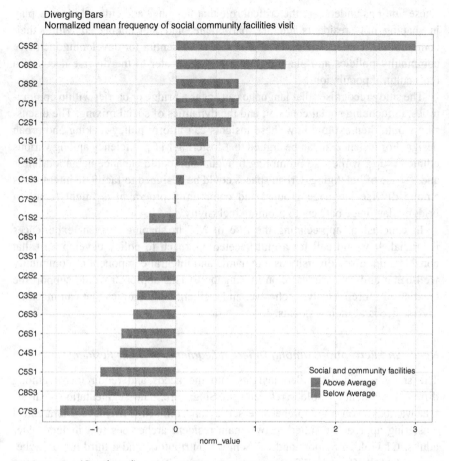

**FIGURE 1.15** (Continued)

$(F(3, 713) = 5.67, p = 0.001)$. Non-English-speaking residents showed higher engagement in community gatherings and group activities, reflecting stronger community ties within these groups.

- *Community Contribution:* The analysis found significant variations in the level of involvement and willingness to help within the community $(F(3, 713) = 3.98, p = 0.008)$. Residents who speak their MT more frequently were more likely to contribute to community activities and volunteer, suggesting that cultural and linguistic background enhances community involvement.

- *Sense of Belonging:* There were marked differences in residents' emotional attachment to their community based on their MT $(F(3, 713) = 6.45, p < 0.001)$. Non-English speakers generally reported a stronger sense of belonging and connection to their community compared to their English-speaking counterparts.

These findings underscore the critical role that language and cultural identity play in shaping how residents interact, participate, and feel connected within their communities. Understanding these dynamics is essential for developing inclusive community policies and planning initiatives that cater to the diverse needs of a multicultural population.

The study recognises that language can act as a bridge or barrier within communities, influencing social cohesion and the dynamics of social mixing. The discussion would then explore how these insights can inform policy-making and urban design. For instance, it can be argued that fostering MT proficiency among various ethnic groups, while also promoting bilingualism, could enhance neighbourliness and a sense of belonging. Urban spaces could be designed to facilitate interactions among diverse language groups, and community programmes might focus on bridging language barriers to promote inclusivity.

In conclusion, appreciating the role of MT in shaping resident experiences is crucial. It would call for a multifaceted approach to policy development that considers linguistic diversity as a dynamic and integral component of community well-being and social integration in Singapore. This approach would support the creation of more inclusive, cohesive, and resilient communities that can meet the diverse needs of their residents.

### Needs and Perceptions among Different Segments of Older Residents

The study provided detailed insights into the needs and requirements among different segments of older residents in Singapore. The segmentation analysis that was based on a comprehensive survey and utilised a two-level hierarchical clustering approach resulted in two main resident archetypes that include older adults: C1 Golden Seniors and C5 Senior Contributors, and a third minor archetype C8 Multi-Gemners. The second-level segmentation also gave rise to seven sub-clusters: C1S1 (senior couples without nearby relatives), C1S2 (senior individuals without nearby relatives), C1S3 (senior couples with nearby relatives), C5S1 (employed younger senior contributors), C5S2 (unemployed/retired older senior contributors), C8S2 (unemployed/retired seniors living with adult children), C8S3 (irregularly employed seniors living with adult children). This highlights the heterogeneous nature of the older adult demography, with a diverse spectrum of lifestyle, language used, neighbourly ties, perception of social change and technology, etc., that needs to be taken into account when making policies and designing housing and urban spaces.

For example, the perception of technology varies significantly between the archetypes. Golden Seniors tend to have lower technological self-efficacy, while Silver Contributors exhibit varying degrees of technological engagement. This difference underlines the need for tailored technology education programmes, with some requiring basic digital skills training and others perhaps benefiting from more advanced technology utilisation courses.

The attitudes towards older adults within all eight resident archetypes are generally positive, but there are nuances. For example, some sub-clusters within the Golden Seniors have moderate views on ageism, indicating the need for public awareness campaigns to foster respect and understanding for the elderly.

The different needs between Golden Seniors and Silver Contributors in various aspects can be summarised as follows:

- *Housing and Accessibility*: Golden Seniors may require more age-friendly housing designs and community infrastructures, while Silver Contributors might benefit from policies that ensure housing security as they transition into retirement.
- *Healthcare*: Both segments would benefit from accessible, affordable healthcare services, but Silver Contributors might also need more proactive health management programmes as they are still in the workforce.
- *Community Engagement*: Social programmes may need to be more targeted, with Golden Seniors possibly requiring more in-person community events, while Silver Contributors might be more engaged in community programmes that offer both physical and virtual participation options.
- *Financial Security*: Golden Seniors often live with lower incomes and hence might need financial assistance or subsidies. Silver Contributors might require support in retirement planning to ensure they do not face financial hardships post-retirement.
- *Education and Employment*: Continued education programmes could be beneficial for Silver Contributors who are still active in the workforce, while Golden Seniors might need more recreational, informal learning opportunities.

In conclusion, addressing the needs and requirements of young-old and old-old residents requires a multifaceted approach that considers the diversity within this population. The Golden Seniors need affordable, accessible services and supportive community structures, and educational initiatives to enhance their digital literacy. The Silver Contributors would benefit from retirement planning services and age-friendly technology solutions.

The broader societal attitudes towards the elderly highlight the importance of developing public awareness campaigns that celebrate the elderly and encourage inclusive attitudes and behaviours. Additionally, as technology becomes increasingly integrated into daily life, it is crucial to ensure that older residents are not left behind and are instead empowered to use technology to improve their lives.

Collaborative efforts involving government, community organisations, and technology providers are vital in developing and implementing targeted initiatives. Such collaborations can ensure that policies are not only inclusive but also adaptable to the evolving needs and preferences of the elderly population, ultimately leading to a society that not only supports but also values its older members.

## ACKNOWLEDGEMENT

Parts of this chapter are based on the final report of 'New Urban Kampung Research Programme' – Chapter 3: Shifting Resident Demographics, also co-authored by Ate Poorthuis, Qingqing Chen, and Dylan Lien. The chapter authors thank the report authors for their contribution.

## Notes

1  This scope was based on the original scope of New Urban Kampung Research Program, which comprises Project 1: Geo-sociographic segmentation, Project 2: HDB Quality of Life, and Project 3: Self-enabled Communities.
2  The four levels of neighbourliness in Singapore context is adapted from Guttman Scale of Neighbouring Relations: (1) Geographical availability; (2) Passive contact; (3) Intentional contact; (4) Mutual trust. See Grannis (2009).

## Bibliography

Alesina, A., & Giuliano, P. (2011). Family ties and political participation. *Journal of the European Economic Association, 9*(5), 817–839. https://doi.org/10.1111/j.1542-4774.2011.01024.x

Barton, J., & Pretty, J.N. (2010). What is the best dose of nature and green exercise for improving mental health? A multi-study analysis. *Environmental Science and Technology, 44*(10), 3947–3955.

Bhat, A. (2014). K-Medoids clustering using partitioning around medoids for performing face recognition. *International Journal of Soft Computing, Mathematics and Control, 3*(3), 1–12. https://doi.org/10.14810/ijscmc.2014.3301

Browning, C. R., Calder, C. A., & Soller, B. (2017). Ecological networks and neighbourhood social organisation. *American Journal of Sociology, 122*(6), 1939–1988.

Chen, K., & Chan, A. H. S. (2014). Gerontechnology acceptance by elderly Hong Kong Chinese: A senior technology acceptance model (STAM). *Ergonomics, 57*(5), 635–652. https://doi.org/10.1080/00140139.2014.895855

Cohrun, S. E. (1994). Understanding and enhancing neighbourhood sense of community. *Journal of Planning Literature, 9*(1), 92–99. https://doi.org/10.1177%2F088541229400900105

Crow, G., Allan, G., & Summers, M. (2002). Neither busybodies nor nobodies: Managing proximity and distance in neighbourly relations. *Sociology, 36*(1), 127–145. https://doi.org/10.1177/0038038502036001007

Delhey, J., & Steckermeier, L. C. (2016). The good life, affluence, and self-reported happiness: Introducing the good life index and debunking two popular myths. *World Development, 88,* 50–66. https://doi.org/10.1016/j.worlddev.2016.07.007

Deloitte. (2019a). The Deloitte global millennial survey 2019. Retrieved from https://www2.deloitte.com/content/dam/Deloitte/global/Documents/About-Deloitte/deloitte-2019-millennial-survey.pdf

Diener, E., Emmons, R. A., Larsen, & R. J., & Griffin, S. (1985). The satisfaction with life scale. *Journal of Personality Assessment, 49*(1), 71–75. https://doi.org/10.1207/s15327752jpa4901_13

Dunn, O.J. (1961). Multiple comparisons among means. *Journal of the American Statistical Association, 56,* 52–64.

Fox, J., & Weisberg, S. (2019). An {R} companion to applied regression, Third Edition. Sage. Retrieved from https://socialsciences.mcmaster.ca/jfox/Books/Companion/

Gower, J. C. (1971). A general coefficient of similarity and some of its properties. *Biometrics, 27*(4), 857. https://doi.org/10.2307/2528823

Grannis, R. (2009). *From the ground up: Translating geography into community through neighbour networks* (pp. 17–27). New Jersey: Princeton University Press.

Hastie, T., & Tibshirani, R. (1995). Generalised additive models for medical research. *Statistical Methods in Medical Research, 4*(3), 187–196. https://doi.org/10.1177 %2F096228029500400302

Henning, C., & Lieberg, M. (1996). Strong ties or weak ties? Neighbourhood networks in a new perspective. *Scandinavian Housing and Planning Research, 13*(1), 3–26.

Hipp, J. R., & Perrin, A. J. (2006). Nested loyalties: Local networks' effects on neighbourhood and community cohesion. *Urban Studies, 43*(13), 2503–2523.

Hudson, R. (2006). Ageing in a public space: The roles and functions of civic engagement. *Generations: Journal of the American Society on Aging, 30*(4), 51–58. Retrieved 2021, June 8, from https://www.jstor.org/stable/26555484

Kassambara, A. (2020). rstatix: Pipe-friendly framework for basic statistical tests. R package version 0.6.0. Retrieved from https://CRAN.R-project.org/package=rstatix

Keogh, A. M., Carfray, A., Andrews, S., Jenkins, J., Longmate, R., Nichols, N., Farina, C., & Wilde, D. (2000). Assessing quality of life in routine clinical practice: A pilot study. *EDTNA-ERCA Journal, 26*(2), 27–30. https://doi.org/10.1111/j.1755-6686.2000. tb00089.x

Kraemer, K. L., & Dedrick, J. (2000). Refining and extending the business model with information technology: Dell Computer Corporation. *The Information Society, 16*(1), 5–21. https://doi.org/10.1080/019722400128293

Kroll, C., & Delhey, J. (2013). A happy nation? Opportunities and challenges of using subjective indicators in policymaking. *Social Indicators Research, 114*(1), 13–28. https:// doi.org/10.1007/s11205-013-0380-1

Lim, D. N. J. Y., & Chong, K. H. (2020, Oct. 9). Rise of virtual community initiatives: O2O community development framework [Conference session]. The 13th Conference of the International Forum of Urbanism (IFoU), monitored from Nanjing, China due to the Covid-19 pandemic.

Lüdecke, D., Makowski, D., Ben-Shachar, M., Patil, I., Waggoner, P., & Wiernik, B. (n.d.). Assessment of regression models performance. Retrieved from https://easystats.github.io/ performance/

Martinson, M. (2006). Opportunities or obligations? Civic engagement and older adults. *Generations: Journal of the American Society on Aging, 30*(4), 59–65. Retrieved from https://www.jstor.org/stable/26555486

Massam, B. H. (2002). Quality of life: Public planning and private living. *Progress in Planning, 58*(3), 141–227. https://doi.org/10.1016/S0305-9006(02)00023-5

MOF. (2019). *Key socio-economic outcomes across cohorts*. Retrieved from https://www. mof.gov.sg/docs/default-source/default-document-library/news-and-publications/ featured-reports/mof-report.pdf?sfvrsn=e07d680a_2

Nussbaum, M., & Sen, A. (Eds.) (1993). The quality of life. Oxford: Oxford University Press.

Perkins D. D., & Long D. A. (2002). Neighbourhood sense of community and social capital. In A. T. Fisher, C. C. Sonn, & B. J. Bishop (Ed.), *Psychological sense of community. The plenum series in social/clinical psychology*. New York: Springer. https://doi.org/10.1007/ 978-1-4615-0719-2_15

Perren, K., Arber, S., & Davidson, K. (2004). Neighbouring in later life: The influence of socio-economic resources, gender and household composition on neighbourly relationships. *Sociology, 38*(5), 965–984. https://doi.org/10.1177/0038038504047181

Phillips, N. D. (2018). Chapter 15: Comparing regression models with ANOVA(). In N. D. Phillips (Ed.), *YaRrr! The pirate's guide to R*. Retrieved from https://bookdown.org/ndphillips/YaRrr/

R Core Team (2020). R: A language and environment for statistical computing. R Foundation for Statistical Computing. Retrieved from https://www.R-project.org/

Ramos-Vidal, I. (2019). A relational view of psychological empowerment and sense of community in academic contexts: A preliminary study. *Behavioural Sciences, 9*(6), 65. https://doi.org/10.3390/bs9060065

Sampson, R. J., Stephen, W. R., & Earls, F. (1997). Neighbourhoods and violent crime: A multilevel study of collective efficacy. *Science, 277*, 918–924. https://doi.org/10.1126/science.277.5328.918

Sirgy, M. J., Reilly, N. P., Wu, J., & Efraty, D. (2012). Review of research related to quality of work life (QWL) programs. In K. C. Land, A. C. Michalos, & M. J. Sirgy (Eds.), *Handbook of social indicators and quality of life research* (pp. 297–311). Dordrecht; Heidelberg; London; New York: Springer Netherlands. https://doi.org/10.1007/978-94-007-2421-1_13

Surrain, S., & Luk, G. (2019). Describing bilinguals: A systematic review of labels and descriptions used in the literature between 2005–2015. *Bilingualism: Language and Cognition, 22*(2), 401–415. https://doi.org/10.1017/S1366728917000682

Töpfer, M., & Bug, P. (2015). *Classical consumer lifestyle segmentation methods*. Reutlingen, Germany: Reutlingen University.

Veenhoven, R. (2002). Why social policy needs subjective indicators. In M. R. Hagerty, J. Vogel, & V. Møller (Eds.), *Assessing quality of life and living conditions to guide national policy* (Vol. 11, pp. 33–45). Kluwer Academic Publishers. https://doi.org/10.1007/0-306-47513-8_3

Wickes, R., Zahnow, R., Corcoran, J., & Hipp, J. R. (2018). Neighbourhood social conduits and resident social cohesion. *Urban Studies, 56*(1), 226–248. https://doi.org/10.1177%2F0042098018780617

Woodcraft, S. B., & Dixon, T. (2013). Creating strong communities–measuring social sustainability in new housing development. *Town and Country Planning Association, 82*(11), Article 11.

Yen, I. H., Michael, Y. L., & Perdue, L. (2009). Neighbourhood environment in studies of health of older adults. *American Journal of Preventive Medicine, 37*(5), 455–463. https://doi.org/10.1016/j.amepre.2009.06.022

Yow, W. Q., & Li, X. (2015). Balanced bilingualism and early age of second language acquisition as the underlying mechanisms of a bilingual executive control advantage: Why variations in bilingual experiences matter. *Frontiers in Psychology, 6*. https://doi.org/10.3389/fpsyg.2015.00164

Zapf, W. (2000). Social reporting in the 1970s and in the 1990s. *Social Indicators Research, 51*(1), 1–15. https://doi.org/10.1023/A:1006997731263

# 2

# NEIGHBOURHOOD QUALITY OF LIFE AND PLACE

*Sin Mei Cheng, Mihye Cho, Jiyoun Kim and Keng Hua Chong*

## Introduction

In October 2019, Singapore's Ministry of Finance released a report comparing socio-economic outcomes across generational cohorts. Compared to the older generations, Singaporeans today are better educated, are more likely to be employed, earn higher wages, and have longer life expectancies (Ministry of Finance, 2019). However, in another survey conducted by Deloitte several months prior, Singapore millennials were more pessimistic about their financial health and the economic climate than their global peers (Seah, 2019). Moreover, only 20% of Singaporean millennials reported being satisfied with their lives. These two findings, although seemingly at odds with each other, and yet valid on their own terms, illustrate the multidimensional challenges of assessing the state of societal progress and the degree to which it enhances individuals' Quality of Life (QoL) and well-being.

Over the last 50 years, the effort by governments and researchers to understand and evaluate social and economic progress has created a plethora of concepts related to QoL such as community well-being, human development, flourishing, liveability, and sustainable development (Massam, 2002; Shek & Wu, 2018). Despite the voluminous, multidisciplinary literature on defining and measuring QoL, its flexible and context-specific nature has made it difficult to tie down a consensus understanding. Notwithstanding their respective nuances, what unites these concepts is the attempt to guide social change and development towards a better society. In short, the notion of QoL is fundamentally about a society's aspirations towards a 'good' life, and how to organise social and economic resources towards such goals. What the 'good' life refers to has evolved through the times, and arises out of the interplay between diverse aspects of human and social organisation. In contemporary terms, the 'good life' encompasses goals such as good health, comfortable and

DOI: 10.4324/9781003437659-4

safe living conditions, economic opportunities and financial security, supportive social environments, cultural richness, political participation, as well as stable and effective physical, legal, and institutional infrastructures to enable and track progress towards such outcomes (Carmona, 2019).

Although this multidimensional nature of QoL seems common-sensical today, the genealogy of the concept shows its co-evolution with the socio-cultural and epistemological milieu of society. The notion of QoL emerged from the social indicators movement during the 1960s to address a growing disenfranchisement with the fixation on infrastructural development, and to incorporate a qualitative element to growth. The economic instability from a crumbling Bretton Woods system, alongside the nascent environmental movement, created an impetus to adopt more expansive notions of growth and progress. As the social indicators movement gained traction in the 1970s, it also benefited from disciplinary contributions from sociology, psychology, economics, political science, and geography. During this time, QoL was stretched from being a largely positivist one to one that encompassed subjective perceptions of individuals' lives, and was embraced by organisations such as OECD and UNESCO (Sirgy et al., 2010). Measures of subjective well-being such as individual traits, aspirations, values, and experiences increasingly complemented 'hard' indicators, and became mainstreamed into research and policy discourse (Veenhoven, 2002).

The 1990s witnessed the further development of objective and subjective indicators to include notions of capabilities and freedoms (Nussbaum & Sen, 1993; Yuan et al., 1999). Such ideas also travelled widely, enabled by transnational knowledge networks such as the UN and the OECD, giving rise to measures such as the Human Development Index which sought to counterbalance national preoccupations with GDP growth. At the same time, globalising economic relations within and between countries gave rise to an emerging awareness of a global marketplace, setting the stage for today's interconnected and hyper-competitive global economic landscape. To this end, QoL and its umbrella concepts were increasingly mobilised towards such competition through a proliferation of comparative QoL studies at national and inter-city levels. While the contemporary construct of 'liveability' was first coined in the context of a Vancouver city initiative to promote equitable planning of neighbourhoods (Ley, 1990; Kaal, 2011), it has since morphed into a rallying concept to highlight how attractive a place is for people to live, work, play, invest, and do business.

By virtue of its fuzzy conceptual boundaries, the notion of QoL has enabled the promotion of various policy agendas. As governments worldwide struggle with growing societal inequalities, polarised publics, and environmental crises, so has the conceptual boundaries of QoL been expanded to reorient governmental objectives (Shek & Wu, 2018). However, applying the QoL concept is challenging because of several reasons. QoL frameworks typically rely on a set of indicators to measure, track, and inform decision-making. These indicators reflect the importance and magnitude of the phenomenon they measure, and can cover a potentially enormous number of social, economic, political, cultural, environmental domains and indicators (Kroll & Delhey, 2013). As such, the selection of domains and indicators to reflect QoL can sometimes seem arbitrary as designers negotiate the tensions between parsimony, accuracy, and availability of data. This makes understanding and measuring QoL a highly context-specific one, as particular cities, regions, and countries possess a unique mix of historical, socio-cultural, economic, political, and environmental characteristics which influences the relevance of specific domains and indicators. Lastly, QoL indicators can be interpreted as both a cause and an effect of well-being, making it crucial to establish valid proxies for the domains being measured. They can be used as a gauge of how attractive a place is, denoting key factors contributing to QoL such as quality of infrastructure, safety, and economic opportunities. At the same time, they can also represent QoL outcomes like life expectancy and satisfaction.

Recently, popularity of QoL indices such as the Economist Intelligence Unit's Liveability Rankings and the Mercer Quality of Living Index have intensified inter-city competition for talent and capital. Such indices apply a standardised set of QoL indicators across cities, ranking each city's attractiveness for people to live, work, play, and invest. Today, there are more than 40 such rankings, and more than 500 indeces and ranking benchmarking cities' performance have been published since 2007 (Jain & Hamel, 2022). This benchmarking and ranking of cities has undeniably heightened the visibility of the liveability agenda. However, it is questionable whether these rankings provide significant utility to policy-makers and urban communities (Berger & Bristow, 2009). Critics have charged that such rankings present an oversimplified view of QoL. Moreover, the utility of such high-profile rankings depends on their intended audiences. While the Mercer and EIU rankings aim to help multinational firms decide on an appropriate compensation package for relocated workers, it may have little utility to policy-makers on improving QoL for ordinary citizens. Indeed, when indicators from the Mercer rankings were juxtaposed with perceived satisfaction based in the same cities, there was weak correlation between the two (Okulicz-Kozaryn, 2013).

Policy-oriented indices such as the Global Liveable Cities Index (GLCI), designed by the Lee Kuan Yew School of Public Policy, aim to approach QoL from the perspective of the ordinary person (LKYSPP, 2012). It calculates QoL from five domains (economic vibrancy and competitiveness, environmental friendliness and sustainability, domestic security and stability, QoL and diversity, and good governance and effective leadership). The GLCI aims to distance itself from the critics of urban branding by striking a balance between drivers of economic competitiveness and the ability of a city to offer a comfortable life to ordinary people. By doing so, the links between effective urban governance and the everyday QoL become clearer to policy-makers. However, the strength of the comprehensive approach of the GLCI may also be its weakness through the index's objective of benchmarking and ranking inter-city differences. This comparative approach, while potentially useful in illustrating generalisable urban knowledge, may overlook highly localised and contextual dynamics of QoL. For instance, GLCI indicators are drawn from mostly objective metrics such as the quality of infrastructure and segment of population with access to various infrastructural services such as internet services and per capita health expenditure. This objective and easily quantifiable bias of the GLCI may facilitate inter-city comparison and benchmarking, but neglects to consider the subjective, intangible, and highly political drivers of well-being in modern cities.

It is likely these tensions between the objective/subjective and global/local in measuring urban performance and QoL will never be completely resolved. For us, our approach to understanding and measuring QoL begins from an individual's lived experiences within and beyond their neighbourhood environment. We argue that examining QoL at the neighbourhood level as intricately linked to Quality of Place (QoP) is useful because peoples' experiences of everyday life provide a more inclusive and textured reflection of policy designs that can complement conventional objective measures. The neighbourhood is not merely for a home but is also a multi-layered space for various functions: (1) a personal space for individual relaxation and familial relationship; (2) an extended space that includes corridors, void decks, and passages which cultivates a sense of community, (3) the landscape surrounding Housing & Development Board (HDB) also provides residents with the spatial and social boundaries that characterise their local living conditions.

## QoL and QoP

An hour and a half long commute to work can rob people of time spent with family or on leisure. Having to navigate a tricky traffic junction on a daily basis can significantly affect day-to-day life for an elderly person. A family of eight having to live in a two-room flat can potentially affect familial relations, lead to disputes with neighbours, and have short and long-term deleterious impacts on physical and mental health. The environment we live in is irrevocably tied to our day-to-day experiences, and, by extension, the dimension of a place and dimension of life of residents are closely tied. The relation between people and place is complex and intricate; each reinforces the other and both are subject to local and global forces of structuration driven by politics, culture, and society. Therefore, to understand and assess the degree to which people are able to lead a 'good life' and the degree to which a place is able to support various groups of people in leading a good life, we need a framework that allows us to take into consideration a range of dimensions that shape everyday life such as health, identity, material wealth, and governance. QoL is a holistic and multidimensional construct that offers us the possibility to develop such a framework.

Broadly defined, QoL is a holistic approach to understanding individual and community well-being, including both objective and subjective indicators. QoL is a complex and multifaceted concept comprising key relational aspects such as individual versus community, person versus environment, aspirations versus achievements, subjective versus objective.

QoL is often constructed by way of domains, sub-domains, and indicators. Domains are large aspects that constitute one's life such as health, material wealth, and social life. Sub-domains are sub-aspects of each domain, for example, for the domain 'material wealth,' sub-domains could include 'jobs and income' and 'consumption.' Indicators are measurable components of domains and sub-domains that are usually derived by taking into consideration availability of data and how they can be tied in with existing policies and/or inform new policies. As stated above, QoL indicators can be subjective and objective in nature; however, they both relate back to an individual. For example, subjective indicators for the sub-domain 'jobs and income' could be perceived satisfaction with one's job and one's income and an objective indicator could be income earned.

In order to retain the holistic nature of the concept, all domains and sub-domains that are of importance for residents should be considered; in order to contextualise these domains to neighbourhood living we also propose the QoP concept to better understand how current and future neighbourhood environments can better support the QoL domains that are of importance to residents. QoP is defined as the measure of factors relating to the physical environment and related social policies that contribute to QoL.

QoP is integral to QoL. While household income, education, health, familial relations all bear importance to the life a person leads, so too do transportation systems that move people between work and home, presence and access to learning and healthcare facilities, and safe places of recreation for individuals and families. QoP is determined by the objective conditions of a place, i.e. accessibility, connectivity, housing, amenities, and facilities, as well as by how these are perceived and utilised by its various residents (subjective notions of place and behavioural patterns). For example, the presence of parks in a neighbourhood and the extent to which residents are satisfied with these parks both contribute to QoP. However, satisfaction with parks and their presence does not entirely determine whether these will be used by residents or not. Working parents might not be able to take their children to the park because their daily routines are overrun by demands from work and home care, or because they prefer their children to play at home or have to keep them home in response to demands from schoolwork. Understanding a resident's perspective from a QoL construct allows for understanding place from other dimensions in people's lives such as work, family, education, and so on. Having QoL and QoP as co-constructs in understanding people and place allows us to see places beyond the technocratic formulae often used to design, plan, and programme places. Therefore, both individual and place-based measures are critical to understanding and assessing quality of neighbourhood life for various groups or segments of the population as well as for specific neighbourhoods.

## Methodology

In our research, a mixed methods approach was taken to generate and assess QoL in public housing neighbourhoods in Singapore. This comprised in-depth interviews, semi-structured interviews with a Q-method[1] component, and a large-scale survey. The description of each phase and method is provided in Figure 2.1.

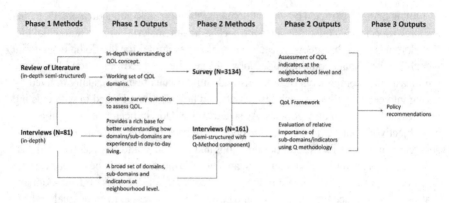

**FIGURE 2.1** Overview of research methods and associated outputs.

*Source*: Authors.

## Phase 1 Interviews

A semi-structured interview guide, with largely open-ended questions, was developed to better understand what matters to residents in leading a good life or a life that they value, and how the living environment supports their everyday living (Table 2.1). In our research, 80 participants were recruited through a range of approaches.[2]

Monthly household income levels of participants ranged from 1,000 to 25,000. The age of participants ranged from 21 to 76, of whom 48 were women and 32 were men. Participants came from diverse backgrounds; we interviewed university students, a range of working adults (such as stay-at-home and working parents, stay-at-home and working elderly citizens, individuals running start-ups or working several jobs at once such as private hire drivers), families living in different types of housing and having a range of educational backgrounds.

Analysis for Phase 1 was conducted in three stages:

- For the first stage of analysis a representative subset of 40 interviews were analysed to develop a coding framework for analysing all interviews and to simultaneously identify relevant indicators for the survey instrument that would align with the coding framework to allow for cross-comparability of data generated by both methods.
- In the second stage of analysis the coding framework developed in the first stage was set up in Atlas Ti (qualitative analysis software) and used to gain an in-depth and detailed understanding of how various domains are manifest in the everyday lives of residents and in what ways their current neighbourhood environment supported their day-to-day living. At a finer level, the coding framework allowed for the distillation of values, attitudes, and behaviours

**TABLE 2.1** Interview guide for Phase 1

*Phase 1 Interview Guide*

General
1   What does the word 'kampung' mean to you?
2   What comes to mind when you hear the word 'kampung'?
3   For older participants:
4   How would you describe your stay at the kampung?
5   Do you remember the configuration of your kampung?
6   How was it different living in a kampung and HDB?
7   For younger participants:
8   Perhaps to make it easier, you can think back on the times you learnt about the
      kampung, what are your general opinions of it?
10  Would you want to experience the kampung if given the chance and why/why not?

Present-day kampung
1   What are some examples of 'kampung' that you can think of in present-day Singapore
      (places, events, etc.)?

Kampung spirit
1   Are you familiar with the terms 'kampung spirit' or 'gotong royong'? What does it
      mean? Please elaborate with some examples.
2   What are your opinions on this term?
3   Where have you heard/seen it being used?
4   What are some of the places in Singapore that encourage 'kampung spirit'?
      Prompt – what about HDBs? Could you please explain? Why or why not?
5   Can you tell me more about your personal experiences of kampung spirit if any?

related to various domains and sub-domains, as well as an understanding of aspirations and future needs of participants related to their QoL. Additionally, each domain and respective sub-domains were closely examined across various levels (such as age, gender, housing type) to derive relevant indicators for QoL.
• Finally, in the third stage interviews were mined to gather support for survey findings at the neighbourhood level and cluster level assessment of QoL.

### Phase 2 Survey and Interview

A large-scale survey was conducted with 3,134 residents living in three selected towns. The survey comprised items derived from Phase 1 interviews to assess and understand QoL. Refer to Chapter 1 for more details on the survey.

A semi-structured interview guide was developed (Table 2.2) and 161 interviews were conducted across the three towns (Toa Payoh N = 58; Jurong East N = 57; Punggol N = 46).

**TABLE 2.2** Interview guide for Phase 2

---

*Phase 2 Interview Guidew*

Daily routines and QoL

1   We are interested in learning more about what matters to you in leading your life. It will be great if you could fill this daily time chart for a typical weekday and weekend day. Please add this sticker to any of the activities that are outside your neighbourhood.
2   What matters to you for leading a life that you value/what is a good life?
3   What are you satisfied with or not satisfied with when it comes to your housing or neighbourhood?

---

Q-sorting

1   Why are particular items ranked higher or lower as compared to others?
2   Why is this particular item ranked the highest or lowest?
3   Are there other items that should be included?

---

Post-sorting questions

1   Why are particular items ranked higher or lower as compared to others?
2   Why is this particular item ranked the highest or lowest?
3   Are there other items that should be included?

---

Changes after 20 years

1   Say, 20 years from now do you think this (point to the Q-sort) would remain the same? If not, what would change and why? Consequently, would your priorities in neighbourhood changes stay the same? If not, what would change and why?

---

Initiatives for neighbourhood

1   Based on your choices above, are there any changes you would like to propose? What are they? Why are they important?
    Examples:
    Co-working spaces in my neighbourhood
    Smart technology/features in homes
    Transforming CCs/RCs into a 'smart' space with more technological features
    More online communication in neighbourhoods
    Pet-friendly homes/neighbourhoods
    Better recycling programmes
    Pay-as-you-throw programme

---

*Source*: Authors.

The Phase 2 interviews were analysed in two stages:

- In Stage 1 R-factor analysis was used to derive five QoL perspectives (Table 2.3) that are manifest in the larger society.
- In Stage 2 a coding framework based on the 49 aspects of QoL derived in Phase 1 interviews was used to gain a more nuanced understanding of QoL aspects

**TABLE 2.3** Five QoL perspectives in Singapore

*Five QoL Perspectives in Singapore*

| Perspective 1 | Perspective 2 | Perspective 3 | Perspective 4 | Perspective 5 |
|---|---|---|---|---|
| *Career and Family Builders* | *A Religious and Meaningful Life* | *Place-Based Community* | *The Aspirational Working Class* | *Healthy and Active Living* |
| Predominantly working-age individuals, 20–49 years old | Mostly middle-class individuals, 20–49 years | Mainly older adults, median age of 63. Over half are >65, retired, and live alone. | Mostly older, less educated, female adults above 40. | Encompasses a diverse group, from ages 21 to 83. Income: $800–$8,000, with various employment statuses. |
| Median household income: $8,000. | Median household income: $6,000. | Median income: $1,800, majority have < $1,000 monthly. | Face challenges of balancing livelihood and caregiving. | |
| Prioritise stable employment and income to support family needs, and personal leisure activities. | Prioritise religion and family as central to personal identity and meaningful life. | Health, affordability, and social support are prioritised, with emphasis on preventive healthcare, affordable living, and strong community ties. | Education, savings, and affordable living are prioritised, driven by a belief in education's social mobility. | Prioritising healthy living and social, emotional, and mental well-being. |
| View government performance positively. | Personal time is valued as well. | | | Engage in activities like healthy eating, exercise, with friends. |
| Issues such as migration, grassroots, and policy-making involvement are less valued. Community engagement is also less prioritised due to busy schedules or other interests. | Career, material wealth, civic participation, and place-based community involvement are less important, with a focus on enjoying life. | Government performance and policy influence are valued due to reliance on state care. | Participants value personal development but struggle with financial insecurity, prioritising necessities over leisure and civic engagement. | Flexible work and social engagements compensate for lack of steady income, hence values job security and affordable living. |

(Continued)

**TABLE 2.3** (Continued)

*Five QoL Perspectives in Singapore*

| Perspective 1 | Perspective 2 | Perspective 3 | Perspective 4 | Perspective 5 |
|---|---|---|---|---|
| "I hardly stay home. My time is all dedicated to work. Because now at this stage, I'm saving as much as possible…" –29 yo, single male resident living with parents in Punggol | "So, these (religion & spirituality) are important for me to be a peaceful, life giving person, positive person. Religion makes a great difference for me." – 49 yo part-time executive and mother of one, living in Punggol | "Especially at my age, healthcare is very important. Of course, being affordable is also good. Because people my age don't have much savings. I need this too, because it is important. Living costs are coming up." – 81 yo single female, retired, living in Toa Payoh | "Owning a house is important. You will feel like you're on the safe side, you have a shelter and education is good to upgrade yourself. And it's a channel to reach other areas." – 74 yo, female, retiree, living in Toa Payoh | "Since we joined as a group…it's very fun… we have a fellowship. But if it coincides with my volunteer programs, then I will go for them because I earn \$30 (from participating)." –65 yo female homemaker, living in Jurong East) |

*Source:* Authors.

across different neighbourhoods. An assessment of QoL at the town, neighbour-hood, and cluster level was conducted using a dashboard tool that visualised survey data, and by looking more closely at interview data from both Phase 1 and 2 interviews. Radar plots were generated at the sub-domain level for each town and cluster. This mixed methods and cross-platform approach allowed for a rich evaluation and elaboration of how various aspects of QoL support every-day life in neighbourhoods at the level of place and cluster.

- To better understand ways through which different individuals perceive different aspects of QoL, we applied the Q-method[1] to collect participants' perception of their QoL across three towns. Each participant was asked to sort 49 Q-statements based on their relative importance to their QoL on a chart resembling a norma-tive distribution. This process allows participants to determine the most and least important aspects of QoL based on their current life situations, using a tan-gible and quantitative medium of expression referred to as a Q-sort (Figure 2.2). Sixty-three out of 99 participants associate significantly with any one of the five perspectives. An overview of each perspective is given in Table 2.3.

A key aspect of this research is to make evident ways in which the physical aspects of homes, neighbourhoods, and towns support everyday life. Hence, the overall coding framework comprises 'Living Environment' as one of the core domains (Table 2.4), while all other domains have been elaborated to signify the intersec-tions of physical and social life where relevant. The act of deriving the framework from our conversations with residents is ultimately a process of abstraction com-prising several interpretive layers. To ensure impartiality in the process, the follow-ing are taken into consideration:

- Organising and interpreting data using first and second cycle coding: Interviews were analysed using first and second cycle coding. First cycle coding involved largely descriptive coding to capture the range of aspects related to QoL. Sec-ond cycle coding involved linking related descriptive codes to larger conceptual codes and adding evaluation coding to scaled aspects of the environment (e.g. flat, block, and neighbourhood). The second cycle coding framework formed a preliminary basis for articulating the QoL framework.
- In-depth and iterative reading of individual narratives: In large-scale qualitative studies such as ours there is an inherent danger to splice individual narratives across categories; to avoid this we often read quotations linked to codes in the larger context of individual interviews to pay attention to personal nuances and contradictions within individual narratives. This process keeps in check norma-tive constructions of concepts that often tend to creep into categorical coding.
- Discussions with peers (both local and non-local): Regular discussions with team mates allowed for tackling individual biases and provided clarity and coherence to the conceptualised abstractions of domains, sub-domains, and indicators. As

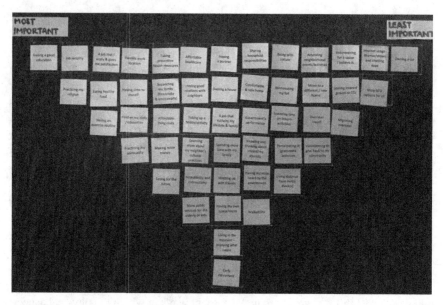

**FIGURE 2.2** Example of a completed Q-sort.

*Source*: Authors.

each member often worked on two to three life domains, team discussions were vital to ensure internal coherence of each domain and the overall framework.[3]

The next section describes in detail the QoL framework and how the QoL and QoP indicators were generated using both interview and survey data and analysis.

## A New Neighbourhood QoL Framework

Indicators, in QoL and related research, are largely generated in a top-down manner, i.e. developed in consultation with experts or derived from existing QoL frameworks, and primarily used to compare QoL in different cities and nations and monitor trends. In recent years, however, some city or neighbourhood scale studies have used bottom-up strategies to develop indicators that reflect local conditions, priorities, and relevance often through partnerships between local governments and citizens groups (Kingsley, 1998; Liu et al., 2023).

Through such bottom-up strategies, our proposed neighbourhood QoL framework is able to encapsulate the material, social, psychological, political, and environmental aspects of neighbourhood living. The framework has been developed through an iterative process where the broad life domains have been modified and redefined by the derivation of sub-domains and indicators, in close alignment with the lived experiences of residents drawn from interviews and surveys of Phase 1 and Phase 2 interviews.

**TABLE 2.4** Example of Phase 1 and 2 coding process

*Life Domain: Living Environment*

| Sub-domain | Indicators | Elaboration of Indicators | Interview Excerpts from Three Towns | | |
| --- | --- | --- | --- | --- | --- |
| | | | *Jurong East* | *Toa Payoh* | *Punggol* |
| Block/ Precinct/ Town | Accessibility to Point Of Interest (POI) | 1 Ease of access to locations/POI within town by mode of transport of choice | 1 Convenience of travelling to Johor Bahru | 1 Very good location in terms of connectivity with other parts of Singapore | 1 MRT is a distance away |
| | | 2 Ease of access to other parts of Singapore by mode of transport of choice | 2 Convenience of travelling to town (Orchard) by express bus service | 2 Very good accessibility, five to ten minutes' walk from home to common points of interest | 2 Peak hour traffic is very bad |
| | | 3 Ease of access to areas beyond Singapore by mode of transport by choice | 3 MRT (Chinese Garden) is a distance away | 3 Not wheelchair friendly | 3 Road safety signages are lacking |
| | | | 4 Not wheelchair friendly | | 4 Punggol is considered isolated from most parts of Singapore |
| | Quality of eating places | 1 Availability of eating facilities in neighbourhood | 1 Numerous options across wide price range | | 1 Pricey food even for coffee shops |
| | | 2 Affordability of eating facilities in neighbourhood | 2 Some of the most famous local foods can be found here, a source of pride for residents | | 2 Limited options due to contained nature of Punggol |
| | | 3 Quality of such eating facilities/food options | | | 3 Prefer to go to Sengkang, Hougang and even towns further away (Toa Payoh) |

*(Continued)*

**TABLE 2.4** (Continued)

*Life Domain: Living Environment*

| Sub-domain | Indicators | Elaboration of Indicators | Interview Excerpts from Three Towns | | |
| --- | --- | --- | --- | --- | --- |
| | | | *Jurong East* | *Toa Payoh* | *Punggol* |
| Natural environment | Proximity to green space | 1 Proximity of green spaces to everyday living<br><br>2 Usage/significance of green spaces to everyday living | 1 Proximity to Chinese Garden is a plus – a refuge from life's stresses | 1 Respondent(s) would like more green spaces e.g. trees, next to their blocks | 1 Proximity to Coney Island is a plus<br><br>2 The Park Connector is well designed for exercise |
| | Air pollution | 1 Source of air pollution and impact on everyday living | 1 Problems with neighbours who smoke and burn incense<br><br>2 Specific to Punggol: smells from nearby petrochemical industrial district bothers current and potential residents | | |

*Source:* Authors.

While the domains within the framework allow for a conceptual understanding of QoL in neighbourhoods, to actually assess and measure its applicability, we need to operationalise each sub-domain. An indicator is an operationalised version of a sub-domain. In most cases a sub-domain can only be fully captured by multiple indicators. Hence, the proposed QoL framework (Figure 2.3) comprises seven main Life domains, 16 Sub-domains, and 67 Indicators (Table 2.5), to comprehensively capture the various aspects of life.

It is also important to note a difference between objective and subjective measures in the framework. Objective measures can be measured in relatively straightforward ways, whereas subjective measures relate to the perception of residents, and thus can often only be measured by asking them directly. Indicators such as Sense of Belonging under the Life Domain of Identity is one such example of subjective measures. Furthermore, we distinguish between QoL (relating to an

**FIGURE 2.3**  HDB QoL framework.

*Source*: Authors.

**TABLE 2.5** Tabulation showing life domains, sub-domains, and indicators

*Tabulation of Life Domains, Sub-domains and Indicators*

| Life Domain | Sub-domain | QoL Indicators | QoP Indicators |
|---|---|---|---|
| Material Wealth | Income | Household Income<br>Income Satisfaction<br>Consumption<br>Affordable Living | Income Inequality<br>Affordable living<br>Economic Vibrancy |
| | Employment | Employment Status<br>Type of Employment<br>Job Satisfaction<br>Short-/Long-Term Unemployment | Employment |
| | Economic Resilience | Personal Assets<br>Perception of Safety Nets<br>Social Safety Nets | Financial Insecurity<br>Social Safety Nets<br>Neighbourhood Change/Stability |
| Social Life | Family Life | Family Time<br>Family-Friendly Culture | Family Time<br>Home-Workplace Proximity |
| | Friendship and Neighbourly Bonds | Quality of Friendships<br>Quality of Neighbourly Ties | Neighbourhood Friendships<br>Neighbourhood Leisure Activities |
| | Leisure and Entertainment | Leisure Time | Quality of Leisure Facilities<br>Inter-neighbourhood Connectivity |
| Living Environment | Mobility and Transport | Public Transport Connectivity<br>Public Transport Affordability<br>Accessibility and Walkability<br>Accessibility of Amenities and Services | Accessibility and Walkability<br>Accessibility to Amenities and Services |
| | Physical and Built Environment | Greenery<br>Noise Pollution<br>Density<br><br>Maintenance<br>Safety and Security | Greenery in Neighbourhood<br>Noise Pollution<br>Neighbourhood Density<br>Neighbourhood Maintenance and Development<br>Safety and Security |

(*Continued*)

**TABLE 2.5** (Continued)

*Tabulation of Life Domains, Sub-domains and Indicators*

| Life Domain | Sub-domain | QoL Indicators | QoP Indicators |
|---|---|---|---|
| | Housing Design | Flat Design | Affordable Housing |
| | | Affordable Living | |
| | | Ventilation | |
| | | Housing Block Design | Urban Design |
| Health and Well-Being | Personal Health and Health Services | (Perceived) Physical Health | Availability of Physical Health Services in the neighbourhood |
| | | Physical Health Services | |
| | | Quality of Physical Health Services | Access to Physical Health Services |
| | | Affordability of Physical Health Services | Neighbourhood Physical |
| | | (Perceived) Mental Health | Health Burden |
| | | Mental Health Services | Perceived Quality of Mental |
| | | Quality of Mental Health Services | Health Services |
| | | Affordability of Mental Health Services | Neighbourhood Mental Health Burden |
| | | | Availability of Mental Health Services |
| | Well-Being and Lifestyle | (Perceived) Stress | Quality of Neighbourhood |
| | | Lifestyle | Lifestyle |
| | | Happiness | |
| Learning | Formal Learning | Education Attainment | Accessibility to School/University |
| | | Affordability of Education | |
| | | (Perceived) Quality of Education | |
| | Informal Learning | Participation in Lifelong Learning | |
| | | Self-Directed Learning | |
| | | Perceived Attitude towards Lifelong Learning | |
| Governance | Trust in Institution | Confidence in Government | Confidence in Town Council's Management |
| | | (Perceived) Governmental Transparency | Perception of Town Council Transparency |
| | Satisfaction with Policy | Housing Policies | |

(*Continued*)

**TABLE 2.5** (Continued)

*Tabulation of Life Domains, Sub-domains and Indicators*

| Life Domain | Sub-domain | QoL Indicators | QoP Indicators |
|---|---|---|---|
| | Civic Engagement | Social Trust<br>Civic Participation<br>Citizenship | Community<br>  participation<br>Civic Participation |
| Identity | Religion | Places of Worship<br>Practising Religion/<br>  Spirituality | |
| | Diversity and<br>Attitudes Towards<br>Differences | Attitudes towards<br>  difference | Neighbourhood<br>  Diversity |
| | Sense of<br>Belonging | Sense of Belonging | Place Attachment |

*Source*: Authors.

individual or family) and QoP (relating to the relationship with the neighbourhood around the individual). QoP indicators can be merely aggregations of QoL indicators. For example, if 'individual employment' is used as a QoL indicator, 'percentage of unemployment in neighbourhood' could be used as a QoP indicator. QoP indicators can also be specific to the neighbourhood scale, without a corresponding QoL indicator. For example, 'diversity of amenities' only makes sense on the neighbourhood level. These form two different dimensions; so a measure can be a potential one out four categories (objective + QoL, subjective + QoL, objective + QoP, subjective + QoP).

Lastly, while the domains provide an analytical framework to understand the various aspects of QoL, they are not unequivocally independent of one another. Indeed, there is much value in examining the relations between domains. Life, after all, is lived and experienced in the intersectionalities of these categories. Also, while life domains and sub-domains are meant to be consistently applicable in the future, the indicators should ideally be revised continuously based on latest insights, policy goals, social issues, and data sources. In the next section, we take a closer look at each individual domain and indicator.

## Life Domains, Sub-domains, and Indicators

### LD1: Material Wealth

Material wealth refers to tangible assets of individuals in a society. It consists of sub-domains of income, employment, and economic resilience.

- Income is a measure of economic resources available to a household to support their overall QoL. Income and other material resources are often gained, and transformed to support one's health, leisure, learning, and other aspects of life, according to each individual's preferences and capabilities. Indicators influencing income include amount of household income, income satisfaction, consumption, and access to affordable living.
- Employment is the main source of income for most Singaporeans, with most expending a considerable amount of time and energy on their careers. Therefore, having work satisfaction can be relevant to one's QoL. Indicators of employment include employment status and type, job satisfaction, and short-/long-term unemployment, all contributing factors in the development of one's social relations, learning, and building of identity.
- Economic resilience refers to the ability to financially recover from or withstand a crisis. Indicators such as personal assets, perception of safety nets, as well as presence of social safety nets are crucial for personal, social, and economic security. They are essential for weathering adverse life circumstances such as losing one's job, sickness, or injury.

### LD2: Social Life

Social life is a crucial aspect contributing to human well-being, providing emotional support, companionship, a sense of belonging essential for mental, emotional, and physical health, as well as personal and societal growth. Family life, friendships and neighbourly bonds, and learning and entertainment are sub-domains influencing social life.

- Families are a foundational unit of society. They provide an individual with the psychological, social, and economic resources for growth and development. Family life indicators include family time and family-friendly culture, which cultivates stable and nurturing family environments vital for social stability and prosperity.
- Friendships and neighbourly bonds are an important source of social and emotional support outside the family unit. Indicators such as the quality of friendships and neighbourly ties help to affirm an individual's place and identity in society, satisfying an individual's need to relate to others and for social connectedness.
- Leisure and entertainment indicators such as spending leisure time contribute to physical and mental well-being, helping one to relax and lead a meaningful life. Leisure activities such as volunteering also promote social cohesion in a community.

## LD3: Living Environment

The living environment of a neighbourhood consists of physical attributes such as the built environment and housing, as well as services such as transportation. These three sub-domains can inform us on the liveability of a place, and, subsequently, the lives of its people.

- Greenery, noise pollution, density, maintenance, safety, and security are indicators of the physical and built environment in Singapore. These are factors most considered by residents with regard to the landscaping and urban build-up of a place.
- Mobility and transport helps us understand the living environment not as a static place but as one of flow and movement. Indicators like connectivity, affordability, walkability, and accessibility both to transport and to amenities in a neighbourhood has the potential to improve one's network and QoL in a given place.
- Housing design plays a significant role in the living environment that determines one's QoL, especially given that one spends at least a third of one's waking hours in the comforts of one's home. Flat design, affordable living, ventilation, housing block design are indicators playing important roles in the development of housing design.

## LD4: Health and Well-Being

A healthy life is perhaps the most basic component of a good life. Personal health and health services as well as wellness and lifestyle are prominent sub-domains of healthy living.

- Personal health, understood in terms of both physical and mental health, is the state of being free from injury or illness, and acts as a good indicator of a society's overall QoL. Thus, a healthy population is one that is more likely to have a healthy appetite for life, and more likely to be actively involved in civic engagements to ensure and maintain their QoL. Pertinent indicators consist of not only one's physical and mental health diagnosis, but also the availability, quality and affordability of both physical and mental health services.
- Well-being goes beyond the narrow terms of physical or mental health, and instead points to a broader, more holistic understanding of one's state of being. Typically, it refers to the state of being comfortable, happy, and healthy. Influencing indicators are perceived stress, lifestyle, and happiness.

## LD5: Learning

Learning helps individuals acquire new knowledge and skills, enabling personal growth and development. Both sub-domains of formal and informal learning are important factors for one's QoL.

- Indicators of formal education are its attainment, affordability, and perceived quality of education. The investment of money and time in higher education is often regarded as an essential prerequisite for a better job, higher salary, and a bright future.
- Informal learning reflects an individual's inclination and capacity for learning beyond formal channels of education. Indicators include participation and perceived attitude towards lifelong learning, as well as self-directed learning, contributing to a broader sense of meaning in life and helping to enrich one's QoL.

### LD6: Governance

Belief in the competence and moral authority of local and national leaders grants legitimacy to the ruling government, and is vital for maintaining order and stability in a society. Sub-domains of governance are trust in institutions, satisfaction with policy, and civic participation.

- Trust in institutions refers to the assurance that citizens have in socio-political institutions to carry out their functions justly, with the belief that the governing bodies are competent, ethical, and capable of addressing the needs of the population. Indicators such as confidence in government and governmental transparency can lead to a knowledgeable and engaged society. When a government allows access to accurate and timely information, it empowers citizens to make informed decisions, which in turn lets them understand pertinent issues and participate meaningfully in civic life.
- Satisfaction with policy indicates the extent to which citizens feel national-level policy is contributing to overall well-being, and is aligned with citizens' aspirations for the country. Indicators such as housing policies can be evident measures of whether citizens have a full understanding of policies that affect well-being, and if they feel their voices are taken into account during policy formulation.
- Civic engagement measures citizens' participation in public affairs. Indicators include social trust, civic participation, and citizenship, measures that can lead to a climate of solidarity, uniting a community with a shared set of values and objectives.

### LD7: Identity

Identity is important in developing self-concept and self-confidence. It is a means through which one develops a healthy sense of self and feels connected to the broader society. Sub-domains such as being comfortable in expressing one's religion, diversity and attitude towards differences, and a sense of belonging are integral to how one makes sense of the world and engages with it.

- Religion imbues individuals with a sense of meaning and purpose, addressing fundamental questions in life. Indicators such as having access to places of worship and the freedom to practise one's religion/spirituality are quintessential to leading a fulfilling life unbridled by biasness or prejudice. Places of worship serve as nodes where like-minded people can gather to support one another, fostering a knitted community of shared values and identity.
- Individuals often align themselves with identities that can be based on one's gender, work, culture, ethnicity, religion, nationality. Indicators such as the diversity and attitude of individuals towards differences in identities can be revealing of how accepting or progressive a society is. An environment that embraces differences allows each individual to live and flourish authentically, and, at the same time, build an inclusive and cohesive society.
- Having a sense of belonging is the feeling of being secure, supported, and connected to a certain group of people or place. Without this, it is easy to feel alone. Belonging is more than simply making social connections, and instead involves building deep meaningful relationships around shared values, beliefs, and goals. Hence, it is intrinsically tied to our life satisfaction, happiness, mental and physical health, forming an important indicator of identity in our lives.

## Applications of the Framework: Town Level

In this section we provide an assessment of QoL in Toa Payoh, Punggol, and Jurong East. Each town-level analysis has an introductory section that provides contextual information about the town. This is followed by a summarised assessment of QoL in each town and, finally, policy recommendations drawn from the assessment.

### Toa Payoh

Located in central Singapore, Toa Payoh was the first town completed by HDB. Because of its long history, its residential population of 121,770 consists of older adults; 53% of survey respondents are above 50 years of age, and 49% belong to single-member households. Most (66%) live in three- or four-room flats. The average household income of survey respondents is $2,500. QoL in Toa Payoh is shaped by its identity as a mature, centrally located town, with great accessibility.

> It is very convenient. Everywhere is reachable within half an hour. If you look at the network of public transport, there is no other place where you can have 2 MRT lines within 10 minutes.
>
> *63, M, Retiree*

However, it also faces issues such as ageing physical infrastructure.

The only thing that I'm not satisfied with is, there's no lifestyle hub here. There's no movie theatre for example. So basically if we want to shop, we need to head down to the mall….so we just accepted it that this part is more of a sleepy estate.

*32, M, Marketing manager*

Despite this, the centrality of the town remains a draw factor to incoming residents, with new residential projects and age-friendly infrastructure leading the town's revitalisation. Figure 2.4 and Table 2.6 illustrate the QoL assessment of Toa Payoh in comparison to the baseline QoL established across all three towns.

The QoL assessment highlights the importance of aligning neighbourhood amenities and facilities with the needs of its population. This suggests a need for a comprehensive approach to town planning and upgrading, including the diversification of amenities and leisure options within ageing neighbourhood spaces, the provision of high-quality public spaces tailored to older adults, increased schooling and childcare options for young families who are moving in, and the enhancement of green spaces for relaxation and improved privacy between buildings.

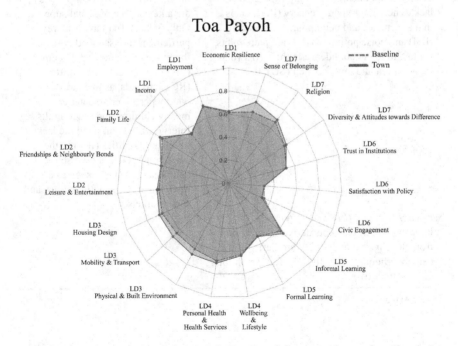

**FIGURE 2.4**   Radar plot visualising QoL in Toa Payoh in comparison to baseline QoL.

*Source*: Authors.

**TABLE 2.6** QoL assessment of Toa Payoh in comparison to baseline QoL

*Toa Payoh*

| *QoL above Baseline* | *QoL below Baseline* |
|---|---|
| *Mobility and Transport (LD3: Living Environment)*<br>Most survey respondents (91%) felt their home is well-connected by transport or walking, scoring high on indicators of inter-neighbourhood and public transport connectivity (IL1, 2), intra-neighbourhood connectivity (IL3), and walkability and accessibility (IL5). | *Income (LD1: Material Wealth)*<br>Public housing affordability (IL4), and whether homes can continue to be viable financial assets during retirement, are of concern, especially to the elderly. |
| *Physical and Built Environment Domain (LD3: Living Environment)*<br>Residents are generally satisfied with flat design (IL1) (87%). Only 20% felt that their house was too small, with the perception that older flats are bigger than new ones. They are also largely content with the urban design of the neighbourhood. However, they expressed a lack of neighbourhood greenery (IL1). Gaps in maintenance and neighbourhood cleaning (IL4) and noise pollution (IL2) are pain points, leading to lower confidence in their grassroots organisation and town council (SD1:IL3). | Civic *Engagement (LD6: Governance)*<br>Most Toa Payoh residents (78%) have good neighbour relations (IL2), although maintained at arms' length. For the elderly, especially those living alone, neighbourhood friendships (IL3) are a key source of social support. Only 36% of Toa Payoh survey participants are satisfied with the range of Community Centre (CC) or Residents' Committee (RC) activities, as they are widely perceived to cater to mostly children and older adults. Interviewees mentioned the lack of leisure facilities (IL3) in the neighbourhood and diverse retail and lifestyle options such as a mall, cinema, speciality stores, and supermarkets. |
| *Sense of Belonging (LD7: Identity)*<br>A large number of residents derive satisfaction from the physical and social familiarity of the neighbourhood, having lived in Toa Payoh for most of their lives. | |

*Source*: Authors.

Additionally, housing design improvements should focus on mitigating residential noise complaints through better enforcement mechanisms, enhancing estate cleanliness by reviewing cleaning processes and implementing education initiatives and mainstreaming rental options to challenge negative stereotypes associated with renters.

Moreover, efforts to enhance neighbourhood leisure and community activities should involve expanding the range of programmes and interest groups to cater to a wider demographic and streamlining the governance of neighbourhood initiatives to encourage greater community participation.

### Punggol

Punggol, a youthful town in Singapore's northeast, was shaped by the "Punggol 21" vision introduced in 1996. Progress faced hurdles due to the Asian financial crisis in the late 1990s, but has since surged. Designed as an 'eco-town' with 'waterfront' living, it blends HDB blocks to resemble private condos. Currently, it houses 138,700 residents, primarily young middle-class families. Despite gradual improvements, concerns exist about slow amenity expansion, public transport connectivity, rising residential density, cost of living, stagnant housing prices, and environmental issues, leaving some residents wanting more supporting infrastructure.

Swimming pool. Because the nearest one (is) at Sengkang... Although (it) is near, but if you take bus, you need to take two (buses). If you take LRT, it's like 45 minutes to get to somewhere so near. So I was thinking how come Punggol is taking so many years to develop things.

*42, F, Homemaker*

I heard from my husband that you will be caught in it (traffic jam) if you do not move out by 7 am. You will be trapped in it if you do not move out by 7 am to 7:30 am. The traffic comes to a standstill in the evening as well.

*39, F, Educator*

We do have a lot of activities for the residents in Punggol ... I mean – but you can see that the locals don't really join. I think they are probably like me, weekend they go out.

*38, F, Homemaker*

Figure 2.5 and Table 2.7 illustrate the QoL assessment of Punggol in comparison to the baseline QoL established across all three towns.

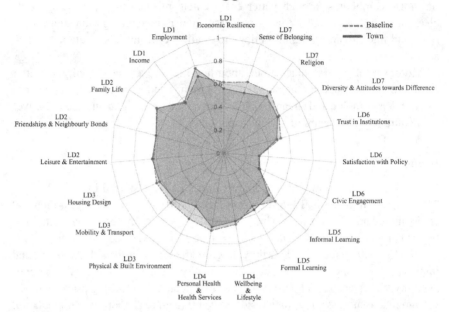

**FIGURE 2.5** Radar plot visualising QoL in Punggol in comparison to baseline QoL.

*Source*: Authors.

The main takeaway from the QoL assessment of Punggol is that supporting physical infrastructure is important to ensure liveability. Key recommendations include synchronising various amenities with housing development to create self-sustained housing estates with proximate facilities.

Housing policies should be inclusive and adaptive for various types of household units, with better communication and additional options for young families seeking to live independently. There is a strong demand for de-stigmatising rental units, and making them available to non-low-income residents.

Housing design should reconsider local community space design to align with residents' needs and routines, promote more socialisation, while dampening noise pollution across homes to accommodate increased time spent at home.

In terms of transport and mobility, measures to ease public transport over crowding and tailor transportation to residents' specific needs, including shuttle buses for estates distant from train stops, and enhancing walkability within and between towns, are essential.

Promoting civic and community participation through participatory activities and resident-led initiatives are also recommended, along with reevaluating the roles of community clubs and resident committees/networks to better align with residents' routines and interests, such as recycling.

**TABLE 2.7** QoL assessment of Punggol in comparison to baseline QoL

| *Punggol* | |
| --- | --- |
| *QoL at and above Baseline* | *QoL below Baseline* |
| Employment (LD1: Material Wealth); and Formal education (LD3: Learning) This is largely due to its young adult demographic. *Social Life (LD2)* 56% are happy with their work-life balance. Similarly, 56% feel they can depend on neighbours for emergencies. | *Living Environment (LD3)* SD1 IL1: Public Transport and Inter-Neighbourhood Connectivity – Punggol is considered well-connected to the city, but not to outlying towns. Bus stops are far away and bus options limited, with overcrowding and traffic jams at peak hour; IL4: Accessibility to Key Amenities and Services scores average (56%), with key amenities such as libraries and public pools lacking. SD2 IL2: Noise from aeroplanes is an issue for 77% of respondents, due to proximity to Seletar Airport; IL4: Some residents (17%) experience bad odours from garbage, raising concerns about maintenance and confidence in the Town Council; IL5 Safety and Security concerns such as remote areas, crimes, and unsanitary rental flat environments were raised. |
| *Satisfaction with Housing Policies (Governance, LD6)* Close to 50% of participants wish for more rental housing options and 77% feel housing options need to be more adaptive to changing society. | *Civic Engagement (Governance, LD6)* Community participation is below average, with only 38% of respondents satisfied with events organised by the RC, and with 23% participation. |

*Source*: Authors.

## Jurong East

Development of Jurong began in the 1970s, by resettlement of nearby villages. Home to 79,600 residents, largely made up of middle-aged residents and lower middle-class families, it has evolved into Singapore's western shopping district, boasting malls like JEM and Westgate today. Since 2017, Jurong has been slated to become Singapore's second Central Business District, forming a regional hub for business, leisure, and tourism. Present-day residents are content with the neighbourhood's QoL, with many (61%) feeling that it has improved. However, they eagerly await improved transport connectivity as plans to decentralise and transform Jurong into a regional centre continue to progress.

Weekends I will jog in this area at the park connectors that links to Jurong Lake Garden. It's very crowded with kids. …Very big with a lot of games for kids. Adults can just go there for strolls and to explore.

*26, M, Student*

Normally we only have like wet market at the other block, but now there are more. Surrounding here nearby, I only go to Jurong Entertainment but now we got JEM, Westgate, Big Box, Jcube. Last time only have one place.

*49, F, Assistant Logistics Manager*

Figure 2.6 and Table 2.8 illustrate the QoL assessment of Punggol in comparison to the baseline QoL established across all three towns.

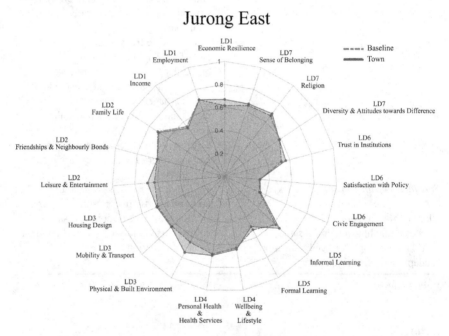

**FIGURE 2.6** Radar plot visualising QoL in Jurong East in comparison to baseline QoL.
*Source*: Authors.

Despite Jurong East's improvements in amenities and services, a more place-based community catering to its residents, especially the elderly population, is needed. Some recommendations include strengthening town council's communications to manage maintenance issues and ensuring common areas are well-maintained for elderly use. Strategies to minimise construction work on weekends and mitigate noise pollution along busy highways are also needed.

**TABLE 2.8** QoL assessment of Jurong East in comparison to baseline QoL

*Jurong East*

| *QoL above Baseline* | *QoL below Baseline* |
| --- | --- |
| *Leisure and Entertainment (LD2: Social Life)*<br>A cluster of new shopping malls, recreational lake area, regional public library, and sports stadium within Jurong East, frequented by families offer space and opportunity for family time and bonding (LD2: SD2: IL2/IL3), improving the perceived QoL (LD2: SD3: IL2/IL3). | *Income (LD1: Material Wealth)*<br>Although many are satisfied with what they earn, there are a sizeable number of lower-income families living in Jurong East. |
| *Physical & Built Environment (LD3: Living Environment)*<br>Neighbourhoods are described as spacious and peaceful, especially with newly developed green spaces such as the Jurong Lake Gardens; 59% of residents expressed satisfaction with the access to green spaces within the neighbourhood (IL1). Indeed, the presence of green spaces buffers the perception of residential density in Jurong East, with only 26% of residents feeling that their houses are too close to one another (IL3). | *Formal Learning (LD5: Learning)*<br>Only 24% of residents in Jurong East have a tertiary education – the lowest percentage across all three towns. But this may be augmented by the high level of satisfaction with the access to facilities and opportunities for Informal Learning. |
| However, although most are happy with redevelopments, some pointed out that new changes have caused earlier common spaces to disappear entirely (SD3: IL4). Local spaces are also underutilised. Forty percent of residents are willing to pay for well-maintained and better quality facilities in the neighbourhood, suggesting a need to reevaluate common spaces design. With the ongoing redevelopments, construction and traffic have also come up as two major sources of noise pollution (IL2). | |
| *Mobility and Transport (LD3: Living Environment)*<br>Although most residents (89%) in Jurong East feel that they are well-connected by public transport to places they regularly go to, the majority wish for better inter-town links to reduce long commutes and transportation costs while travelling out for work or leisure. | |

(*Continued*)

TABLE 2.8 (Continued)

| *Jurong East* | |
| --- | --- |
| *QoL above Baseline* | *QoL below Baseline* |
| *Sense of Belonging (LD7: Identity)*<br>A third of residents are aged 60 and above, hence Accessibility and Walkability is an important factor; 83% of residents believe their neighbourhood is inclusive for all, with ramps enhancing mobility for seniors, resulting in a strong sense of belonging. Most associate their neighbourhood with fond memories (70%), and only 20% plan to move in the next five years. Residents generally have good relationships with neighbours (79%), with many considering them as friends (66%). | |

*Source*: Authors.

In terms of mobility and transport, diversifying bus service types could facilitate efficient inter-town travel and alleviate congestion at the bus interchange. To foster civic and community participation, it was suggested by residents to build on Jurong East's heritage in future planning, such as involving residents in participatory workshops to create green social spaces rooted in its swamp history.

Income assistance measures should focus on maintaining the affordability of food and groceries, and ensuring accessibility to health and leisure facilities.

## Application of QoL within Neighbourhoods

In addition to the application at the town level, the neighbourhood QoL framework can also be applied on a smaller scale at the neighbourhood level, as demonstrated here through the comparative analysis of QoL across two neighbourhoods in Jurong East.

### Jurong East: Comparison of Neighbourhood N3a and Neighbourhood N4a

The radar plots in Figure 2.7 show how the two neighbourhoods in Jurong East perform across all 19 sub-domains of QoL. While the two neighbourhoods are similarly geographically further away from the central area of Jurong East, N3a and N4a perform very differently across many of the QoL sub-domains.

Based on 194 responses from N3a and 218 responses from N4a, N3a demonstrates a need for improvements in almost all sub-domains when compared to the town's QoL, whereas N4a performs above average for most sub-domains.

Potential areas of improvements in N3a include Economic Resilience (LD1), Physical and Built Environment (LD3), and Sense of Belonging (LD7). N4a, on the other hand, surpasses the town's QoL in these three sub-domains. The observed variations between the two neighbourhoods can be attributed to the differing socio-economic statuses of their residents. In N3a, a substantial majority consists of lower-income and single-parent families who rent their housing units directly from HDB. Notably, 74% of residents in N3a earn below the town's median household income of $3,000. Conversely, N4a stands out with a majority of homeowners, where 62% of residents comfortably exceed the town's median household income (Figure 2.8).

This contrast in socio-economic profiles can be further elucidated by examining the employment situations within each neighbourhood (Figure 2.8). In N3a, there is an above-average percentage of residents engaged in precarious work, including part-time jobs (10%) and freelancing (6%). On the other hand, N4a exhibits a different scenario, where a notable proportion of residents work in shifts (11%) or over the weekend (22%), but these are typically regular jobs.

This nuanced understanding of the employment landscape sheds light on the disparities in Economic Resilience (LD1) observed between the two neighbourhoods. The positive employment stability and higher income levels in N4a contribute to its enhanced performance in this domain, highlighting the importance of considering socio-economic factors in assessing neighbourhood dynamics.

A significant 58% majority of residents in N4a are confident that their neighbourhood has experienced positive changes. On the contrary, residents in N3a are evenly split between perceiving their neighbourhood as remaining the same (49%) or improving (46%). A minor percentage (5%) perceives a decline, albeit slightly higher than the town average (3%) (Figure 2.9a).

It is also imperative to look into the perception of noise pollution between the two neighbourhoods. In general, more residents in N4a report no noise pollution (63%) compared to N3a (36%). Residents in both neighbourhoods point to their neighbours as a consistent source of disturbance, with N3a attributing a higher percentage to their neighbours (29%) compared to residents in N4a (11%) (Figure 2.9b).

Residents in both N3a and N4a are relatively new to their neighbourhoods compared to residents staying in other neighbourhoods in Jurong East; 76% of residents in N3a and 61% of residents in N4a have lived in the neighbourhood for less than ten years. It is worth noting that residents in N3a are largely lower-income renters (57%) and therefore constitute a fairly transient neighbourhood (Figure 2.9c). This explains why N3a fared below average when it came to Sense of Belonging (LD7) while N4a was above average.

The analysis above demonstrates the effective application of the HDB QoL framework in uncovering valuable insights not only at the town level but also at the neighbourhood level. It serves as a compelling example that even within geographically similar areas, the needs of a population may vary due to distinct

(a)

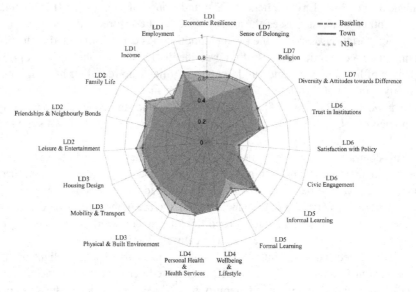

(b)

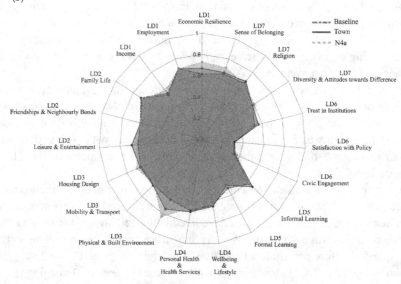

**FIGURE 2.7** Comparison of QoL between N3a and N4a neighbourhoods. The three different layers in the plots visualise baseline QoL, QoL at the town level, and QoL at the neighbourhood level.

*Source*: Authors.

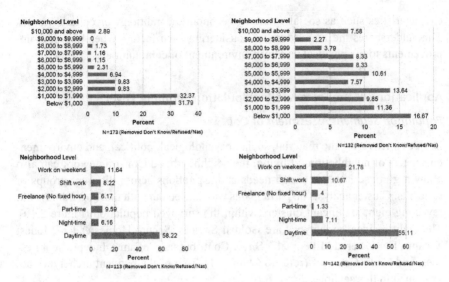

**FIGURE 2.8** Comparison of household income and job time between N3a and N4a.

*Source*: Authors.

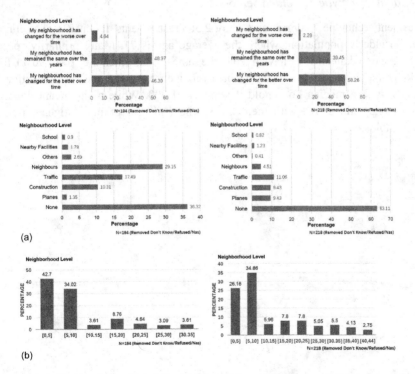

**FIGURE 2.9** Comparison of (a) statement describing neighbourhood, (b) sources of environmental disturbance, and (c) length of residency between N3a and N4a.

*Source*: Authors.

characteristics such as social dynamics, economic conditions, or cultural norms. This underscores the importance of considering such factors when planning improvements to the physical and built environment of communities.

## Applications of the Framework: Population Segment Level

### Population Segment Assessment of QoL

Besides uncovering the material, social, psychological, political, and environmental aspects of neighbourhood living, the neighbourhood QoL framework can also allow a glimpse into the varied needs and aspirations across different groups of residents. Cluster analysis on participants was carried out with the QoL framework, revealing eight significant clusters within the surveyed population (Figure 2.10). The eight resident archetypes are Golden Seniors, Secure Homesteader, Modest Tenants, Empowered MillennialZ, Silver Contributors, CoFam+, Independent Lessees, and Multi-Gemners (refer to Chapter 1). Two of the resident archetypes are examined in this section.

### Resident Archetype 1: Golden Seniors

Resident archetype 1 comprises 14% of 2,568 respondents. It consists of a retired (93%) elderly population, where the average age is 71. Many primarily speak Chinese or a dialect. They have lived in the same house, which they own (90%), for a long time (average 24 years). They have relatively lower incomes and educational levels, with a median household income of $1,000. While a significant portion live alone, others live with their spouse (no children) or in multigenerational households.

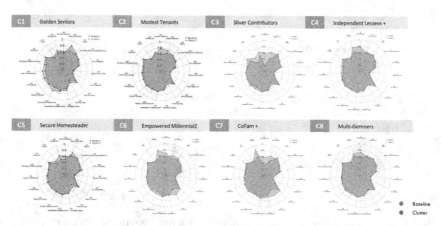

**FIGURE 2.10**   QoL assessment of eight resident archetypes.

*Source*: Authors.

This archetype prioritises health, neighbourhood accessibility, and walkability. Affordable living is important to them, as they are mostly economically vulnerable. They are also socially vulnerable, as most live alone. Although this archetype performs below the baseline in economic domains such as income and employment, it scores above the baseline for most other domains. It scores highly for Sense of Belonging, due to familiarity and attachment to their neighbourhood where they lived for a long time. They also score well for sentiments regarding the Physical and Built Environment and Leisure and Entertainment.

Golden Seniors perform below the average for satisfaction with institutions and policies. Almost half of the participants (46.5%) come from Toa Payoh where data highlights a constructive desire for improved maintenance and cleanliness standards. While 58% of participants acknowledge their grassroots leaders for connecting with the community, they wish for further engagement to strengthen the communication for a more robust representation. Golden Seniors also hope for more improvements in public housing policy, where 74% feel that there should be more housing options for varied household types.

> I do have people who are home-bound, I do marketing for them even though I'm not so well. I have walking difficulties, I help people who are worse than me. For those using the wheelchairs, and also people who are using the walking stick. I help them withdraw money from the bank and cook and pass food. I also help them settle their utility bills, and also help them call an ambulance because most people here stay in a studio apartment.
>
> *74, F, Retiree*

### Resident Archetype 4: Empowered MillennialZ

Resident Archetype 4 comprises 9% of 2,568 respondents. The cluster is young, with an average age of 32 years, and a large percentage of Chinese speakers. Many are employed (64%) but a significant portion is studying full-time (21%). It is largely made up of household types consisting of a couple with adult children (54%), nuclear-plus households (24%), and single-person households (18%). The respondents are mostly adult children, having joined the workforce recently or will do so soon. Many are polytechnic or university degree holders (73%). Concentration of this archetype is higher in Toa Payoh and Jurong East.

Although 91% have close ties with family, empowered MillennialZ is below baseline for family life, as they spend less leisure time with family, preferring friends or partners. The friendships and leisure domain is also below baseline (56% compared to baseline 61%), highlighting fewer neighbourly ties and challenges in meeting friends and spending leisure time.

Empowered MillennialZ is fairly dependent on their family, hence fewer (23% as compared to baseline 34%) feel burdened with household responsibilities. However, only 43% (as compared to baseline 51%) feel that they exercise enough and

54% of employed residents are satisfied with work-life balance (compared to 61% baseline). Despite stress levels being comparable to the baseline, fewer respondents (81% compared to 88% baseline) feel they are in good mental health, possibly because the younger generation associates less stigma with discussing mental health issues.

Most feel safe and secure in their homes; however, they are less satisfied (79% as compared to the baseline 85%) with their physical homes – 27% as compared to 21% baseline feel there are not enough housing options, 50% feel there should be more rental housing options, while 83% (as compared to 79% baseline) feel public housing should cater to more diversity.

> Because all my friends are staying around this area. Let's say I have some trouble or if I feel down during work, sometimes I will meet up with them to have a conversation. Because my friend knows me better than my neighbour, so we have a chat and dinner. That definitely helps to make my work better.
>
> *29 yrs, F*

## Conclusion

The importance of the QoL framework lies in its comprehensive assessment of the well-being and satisfaction of individuals and communities across various dimensions. Improving QoL is essential for creating thriving, inclusive, and resilient societies that prioritise the well-being of individuals and communities. Additionally, the QoP concept is introduced, as a measurement of factors in the physical environment and associated social policies that contribute to QoL.

The utilisation of the QoL framework in the three neighbourhood towns of Toa Payoh, Jurong East, and Punggol serves as a pivotal example in the assessment of neighbourhood living in the local context of Singapore. Conducting comparative studies of neighbourhoods and resident archetypes with the QoL framework allows us to systematically and closely examine various factors impacting residents' lives, including infrastructure, amenities, and social aspects. These studies help draw valuable parallels between different neighbourhoods and resident archetypes, enabling the identification of best practices and successful models that can be applied in future. Insights and recommendations derived from these studies are also beneficial to future phases of works that involve actively collaborating with residents through participatory design and community enablement initiatives. By engaging residents in the decision-making process, their unique perspectives and preferences are at the forefront of the next phase.

This approach is not only beneficial for the local context of the three selected towns but also holds the potential to be generalised to other local towns, or even to international cities. The transferability of insights and QoL framework methodology ensures that successful strategies can be adapted and implemented in diverse urban settings, fostering sustainable and community-centric development on a broader scale.

## Acknowledgement

Parts of this chapter are based on the final report of "New Urban Kampung Research Programme" – Chapter 4: Neighbourhood-Based Quality of Life, also co-authored by Ate Poorthuis, Anupama Reddy, and Qingqing Chen. The chapter authors thank the report authors for their contribution.

## Notes

1 Q-method, or Q-methodology, is a research method developed in the 1930s by William Stephenson, a physicist and psychologist, to study complex issues of human subjectivity. It allows the researcher to conduct a structured and reproducible analysis to understand collective views on a topic, which may include subtle and nuanced differences in opinions.
2 The range of approaches were social media ads, bus-stop ads, snowball and door-to-door interviews. The interviews were conducted at a place convenient for both participants and interviewers. Interviews were recorded with consent from participants. When consent to audio-recording was not given, detailed notes were taken. Interviews on average lasted 50 minutes. Participants were compensated for their time and effort with a $30 grocery voucher.
3 For example in an early iteration of the framework we had the life domains of Family, Social life, and Neighbourliness, however after analysing both phase 1 and 2 interviews we found that for some residents friends and family were interchangeable and in some cases neighbours were friends. To avoid confounding these domains we found having a single life domain for Social Life and sub-domains Family Life, Friendships and Neighbourly Bonds, and Leisure and Entertainment to better capture the inter-relatedness of perceived and lived everyday life.

## Bibliography

Alexander, J. (1990). The sacred and profane information machine: Discourse about the computer as ideology. *Archives de Sciences Sociales Des Religions, 69*, 161–171.
Angelidou, M. (2015). Smart cities: A conjuncture of four forces. *Cities, 47*, 95–106.
Anthopoulos, L. (2017). Smart utopia vs smart reality: Learning by experience from 10 smart city cases. *Cities, 63*, 128–148.
Arun, M., & Yap, M. T. (2000). Singapore: The development of an Intelligent Island and social dividends of information technology. *Urban Studies, 37*(10), 1749–1756.
Barns, S. (2018). Smart cities and urban data platforms: Designing interfaces for smart governance. *City, Culture and Society, 12*, 5–12.
Batty, M. (2012). Smart cities, big data. *Environment and Planning B: Planning and Design, 39*(2), 191–193.
Berger, T., & Bristow, G. (2009). Competitiveness and the benchmarking of nations—a critical reflection. *International Advances in Economic Research, 15*(4), 378–392.
Brenner, N. (2013). *Implosions/explosions: Towards a study of planetary urbanization*. Berlin: Jovis.
Campbell, A., Converse, P. E., & Rodgers, W. L. (1976). *The quality of American life: Perceptions, evaluations, and satisfactions*. New York: Russell Sage Foundation.
Carmona, M. (2019). Principles for public space design: Planning to do better. *Urban Design International, 24*, 47–59. https://doi.org/10.1057/s41289-018-0070-3

Castells, M. (2005). Space of flows, space of places: Materials for a theory of urbanism in the information age. In B. Sanyal (Ed.), *Comparative planning culture* (pp. 45–63). New York, London: Routledge.

Choo, C. W. (1997). IT2000: Singapore's vision of an intelligent island. In P. Droege (Ed.), *Intelligent environments: Spatial aspects of the information revolution* (pp. 49–66). Amsterdam: Elsevier Science.

Featherstone, M. (1990). *Consumer culture and postmodernism.* Published in association with Theory, Culture & Society. LA; London; New Delhi; Singapore: SAGE Publications.

Fredericks, J., Hespanhol, L., Parker, C., et al. (2018). Blending pop-up urbanism and participatory technologies: Challenges and opportunities for inclusive city making. *City, Culture and Society, 12*, 44–53.

Giddens, A. (1981). A contemporary critique of historical materialism. Berkeley: University of California Press.

Glasmeier, A., & Christopherson, S. (2015). Thinking about smart cities. *Cambridge Journal of Regions, Economy and Society, 8*, 3–12.

Granier, B., & Kudo, H. (2016). How are citizens involved in smart cities? Analyzing citizen participation in Japanese 'smart communities'. *Information Polity, 21*, 61–76.

Hatuka, T., & Zur, H. (2020). Who is the 'smart' resident in the digital age? The varied profiles of users and non-users in the contemporary city. *Urban Studies, 57*(6), 1260–1283.

Ho, E. (2017). Smart subjects for a smart nation? Governing (smart) mentalities in Singapore. *Urban Studies, 54*(13), 3101–3118.

Hollands, R. G. (2008). Will the real smart city please stand up?: Intelligent, progressive or entrepreneurial? *City, 12*(3), 303–320.

Housing & Development Board (HDB). (2019). *HDB annual report 2019: Key statistics.* Housing & Development Board (HDB).

Jain, D., & Hamel, P. (2022). *Better rankings for better cities: The limitations and prospects of city rankings.* Retrieved 2024, August 24, from https://www.thenatureofcities.com/2022/04/14/better-rankings-for-better-cities-the-limitations-and-prospects-of-city-rankings/

Kaal, H. (2011). A conceptual history of livability. *City: Analysis of Urban Trends, Culture, Theory, Policy, Action, 15*(5), 532–547. https://doi.org/10.1080/13604813.2011.595094

Kingsley, G. T. (1998). *Neighborhood indicators: Taking advantage of the new potential.* National Neighborhood Indicators Partnership (NNIP), The Urban Institute. https://www.urban.org/sites/default/files/publication/27466/412380-Neighborhood-Indicators-Taking-Advantage-of-the-New-Potential.PDF

Kitchin, R. (2014). The real-time city? Big data and smart urbanism. *GeoJournal, 79*, 1–14.

Kitchin, R., & McArdle, G. (2016). What makes Big Data, Big Data? Exploring the ontological characteristics of 26 datasets. *Big Data & Society, 3*(1), 1–10.

Komninos, N. (2006). Smart cities: Innovation systems and information technologies for urban development. *Architects, 1*(60), 72–76.

Kroll, C., & Delhey, J. A. (2013). Happy nation? Opportunities and challenges of using subjective indicators in policymaking. *Social Indicator Research, 114*, 13–28. https://doi.org/10.1007/s11205-013-0380-1

Lee Kuan Yew School of Public Policy (LKYSPP). (2012). *Launch of global liveable cities index.* https://lkyspp.nus.edu.sg/news-events/events/details/launch-of-global-liveable-cities-index

Ley, D. (1990). Urban liveability in context. *Urban Geography, 11*(1), 31–35. https://doi.org/10.2747/0272-3638.11.1.31

Liu, S., Ge, J., Ye, X., Wu, C., & Bai, M. (2023). Urban vitality assessment at the neighborhood scale with geo-data: A review toward implementation. *Journal of Geographical Sciences, 33*, 1482–1504. https://doi.org/10.1007/s11442-023-2139-1

Logan, J. R., & Molotch, H. (1987). *Urban fortunes: The political economy of place.* Berkeley, Los Angeles: University of California Press.

Luque, A., & Marvin, S. (2015). Developing a critical understanding of smart urbanism? *Urban Studies, 52*(12), 2105–2116.

Massam, B. (2002). Quality of life: Public planning and private living. *Progress in Planning, 58*, 141–227.

Ministry of Finance Singapore. (2019, October). *Key socio-economic outcomes across cohorts.* Ministry of Finance Singapore. Retrieved 2024, August 23, from https://www.mof.gov.sg/docs/default-source/default-document-library/news-and-publications/featured-reports/mof-report.pdf

Nallari, A., & Poorthuis, A. (2021). Rethinking 'kampung' or 'village'in the (re)making of Singapore and Singaporeans. *Singapore Journal of Tropical Geography, 42*(3), 431–450.

Nussbaum, M., & Sen, A. (Eds.). (1993). *The quality of life.* Clarendon Press. https://doi.org/10.1093/0198287976.001.0001

Okulicz-Kozaryn, A. (2013). City life: Rankings (livability) versus perceptions (satisfaction). *Social Indicators Research, 110*(2), 433–451. https://doi.org/10.1007/s11205-011-9939-x

Seah, S. (2019, May 22). Singapore millennials more pessimistic, less satisfied with lives than peers around the world. *Today Online.* https://www.todayonline.com/singapore/singapore-millennials-more-pessimistic-less-satisfied-lives-peers-around-world

Seik, F. T. (2000). Subjective assessment of urban quality of life in Singapore (1997–1998). *Habitat International, 24*(1), 31–49.

Shek, D., & Wu, F. (2018). The social indicators movement: Progress, paradigms, puzzles, promise, and potential research directions. *Social Indicators Research, 135*, 381–406. https://doi.org/10.1007/s11205-017-1552-1

Shelton, T., Zook, M., & Wiig, A. (2015). The 'actually existing smart city'. *Cambridge Journal of Regions, Economy and Society, 8*, 13–25.

Sirgy, M. J., Widgert, R. N., Lee, D.-J., & Yu, G. (2010). Developing a measure of community well-being based on perceptions of impact in various life domains. *Social Indicators Research, 96*, 295–311. https://doi.org/10.1007/s11205-009-9479-9

Teo, S. E., & Kong, L. (1997). Public housing in Singapore: Interpreting "quality" in the 1990s. *Urban Studies, 34*(3), 441–452.

United Nations (2019). *The age of digital interdependence: Report of the UN Secretary-General's high-level panel on digital cooperation.* New York: United Nations Digital Library.

Urry, J. (2000). *Sociology beyond societies: Mobilities for the twenty-first century.* London, New York: Routledge.

Vanolo, A. (2014). Smartmentality: The smart city as disciplinary strategy. *Urban Studies, 51*(5), 883–898.

Veenhoven, R. (2002). Why social policy needs subjective indicators. *Social Indicators Research, 58*(1/3), 33–45. http://www.jstor.org/stable/27527001

Wright, J. D. (1982). The quality of life: Comparative studies by Alexander Szalai and Frank M. Andrews. *Contemporary Sociology, 11*(1), 78–79.

Yeoh, B. (2010). Kampung. In R. Hutchison (Ed.), *Encyclopedia of urban studies* (p. 420). LA; London; New Delhi; Singapore; Washington DC: Sage.

Yuan, L. L., Yuen, B., & Low, C. (Eds.). (1999). *Urban quality of life: Critical issues and options*. NUS Press. https://books.google.com.sg/books?id=_wvIi0O-JH8C&source=gbs_navlinks_s

Zandbergen, D., & Uitermark, J. (2019). In search of the smart citizen: Republican and cybernetic citizenship in the smart city. *Urban Studies*, *57*(8), 1733–1748.

Zukin, S. (1996). *The cultures of cities*. Hoboken, New Jersey: Wiley-Blackwell.

# PART II

# Envisioning Opportunities

By consolidating multi-modal data and uncovering the interdependencies between urban systems and social factors, four urban analytics tools and processes have been developed that can aid socio-environmental data-driven planning and help to identify intervention sites, targeted residents, and areas of opportunities to better residents' quality of life.

DOI: 10.4324/9781003437659-5

# 3
# DATA ANALYTICS FOR COMMUNITY AND PLANNING

*Tshui Mum Ha, Yohei Kato and Keng Hua Chong*

## Introduction

Housing & Development Board (HDB) introduced the 'New Town Structural Model' in the 1970s with the aim to improve public housing residents' Quality of Life (QoL) through the planning and design of the built environment. The model emphasised on spatial and programmatic planning of neighbourhoods by creating a systematic network of places as connected settings to support residents' daily necessities, social interactions, neighbourly bonding, as well as recreational activities (Housing & Development Board, 1995; Ooi & Tan, 1992).

The introduction of the 'precinct' concept in the 1980s and its subsequent evolutions showed the government's continuous effort in addressing the changing needs of the population through physical planning and design. These evolving precinct design ideologies also reflect shifting national interest priorities, especially in the context of social architecture to improve residents' QoL. This can be classified into different stages based on a qualitative comparative analysis, namely, 'collectivism (1977–1990)', 'heartland (1990–1995)', 'kampung spirit (also known as community spirit) (1995–2000)', and 'social capital and sustainability (2000–2015)' (Chong & Tan, 2015).

Until today, the precinct concept remains the most granular level of town planning in Singapore, where a neighbourhood is further subdivided into smaller parcels of social and physical spaces, potentially with distinctive aesthetics. Each precinct is planned to consist of residential blocks housing 400–600 dwelling units (though it varies substantially in reality depending on site conditions), arranged around a central open space that includes common facilities. The precinct concept guides the design of finer details such as street furniture, covered linkways, playgrounds, fitness corners, and so forth, aiming to foster a sense of belonging,

DOI: 10.4324/9781003437659-6

encourage community interaction, and promote bonding within a more intimate setting (Heng, 2017). The consistent efforts by the government to enhance residents' lives through physical planning are seen to reflect the "belief held by planners and architects: the physical environment, in which residents live, shapes and even determines their quality of life and social life" (Dempsey, 2009, pp. 316–317).

But how can we measure the effects of precinct design on improving residents' QoL? How do we ensure a thorough and objective measurement when both precinct design (defined by planning and design parameters) and QoL are multi-faceted? This chapter identifies the need for a more in-depth evaluation of precinct design and its influence on residents' perceived QoL, with two objectives: The first is to discuss the use of multi-dimensional classification (MDC) in identifying precinct typologies based on their planning and design parameters. The second is to apply the said technique in three HDB residential towns in Singapore and, through statistical analysis, examine whether residents living in precincts of different typologies exhibit different levels of perceived QoL.

## Conceptual Framework towards a MDC in the Study of Precinct Design and QoL

MDC refers to the categorisation of items using multiple sets of class variables (Gil-Begue et al., 2021); it usually involves dependencies between class variables (Read et al., 2014). It is essentially a data classification and reduction methodology that creates different categories where the correlations between sub-categories can be drawn with ease. Given the increasing complexity inherent in urban environments where many elements are intertwined, MDC has proven valuable in identifying combined factors that come into play to shape urban living.

For instance, Alexiou et al. (2016) identified different physical built environment characteristics (e.g., housing type, population, buildings, road network, registered parks and gardens, retail centres, etc.) to create an MDC for urban morphologies and understand their relationship to socio-economic profiles. Similarly, Ibes (2015) developed a multi-dimensional urban park classification methodology based on spatial attributes such as physical and spatial park characteristics, land cover, built context, etc. The resulting park types were used to compare to the neighbourhood social characteristics, providing insights into social equality in access to urban parks.

While there is a widespread consensus on the application of MDC in understanding the built environment and urban living, there is a lack of empirical evidence in examining the potential application of this technique at the precinct level, the most granular level in Singapore's public housing context where the vibrancy of civic life and QoL truly manifest, along with precinct facilities that cater to residents' day-to-day needs. Precinct facilities such as playground, fitness corner, and community garden are vital hubs where residents gather for physical and recreational activities, serving as pivotal communal spaces that nurture social interactions.

The development of these facilities within a precinct is guided by distinct design concepts, contributing to the reinforcement of the precinct's unique identity and becoming an integral part of residents' social memories. At this level of planning, the pedestrian network is thoughtfully designed to enhance inter-neighbourhood connectivity and walkability (Housing & Development Board, 2018). Covered linkways are strategically integrated not only to facilitate residents' journey to nearby precinct facilities, but also to create opportunities for chance encounters with neighbours during daily commutes, fostering spontaneous social interactions. These chance encounters and spontaneous social interactions are anticipated to lay the foundation for deeper and more meaningful interactions over time. It is also within this level of planning that the precinct's density is determined, shaping the overall high-rise, high-density urban living experience.

A conceptual framework to explore the relationships between precinct design and residents' perceived QoL using MDC was developed with the following four key questions (Figure 3.1). Each of these key questions serves as a guiding element for the quantitative workflow.

1  What are the planning parameters that guide the design of public housing at the precinct level?
2  How can we identify and cluster different precinct typologies according to their planning parameters?
3  What are the perceived QoL indicators that can be potentially influenced by planning parameters?
4  Do residents living in precincts of different typologies exhibit different levels of QoL?

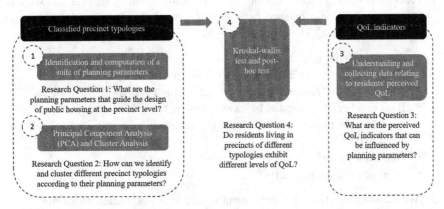

**FIGURE 3.1**   Conceptual framework and methodology to apply multi-dimensional technique in studying the relationships between precinct design and residents' perceived QoL.

*Source:* Authors.

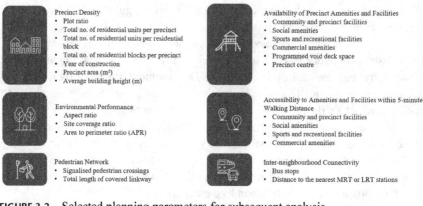

**FIGURE 3.2** Selected planning parameters for subsequent analysis.

*Source:* Authors.

## Identification and Computation of a Suite of Planning Parameters

The identification of planning parameters was conducted based on two main criteria. First, the selected parameters should be in accordance with the HDB planning guidelines, particularly at the precinct level, ensuring the findings can provide practical insights into design policy recommendation. A literature review of the HDB Design Handbook and other planning guidelines was carried out to identify the planning elements that are used to distinctively characterise precinct design. Second, the parameters should be measurable and capable of being operationalised into indicators. The selected planning parameters are detailed below (Figure 3.2).

1 Precinct density: Precinct density pertains to the intensity of land use within a precinct, striking a balance between accommodating housing development and preserving QoL. The measurable indicators include:

- Plot ratio
- Total number of residential units per precinct
- Total number of residential units per residential block
- Total number of residential blocks per precinct
- Year of construction (indicative of the historical planning that influenced housing density)
- Precinct area (m²)
- Average building height (m)

2 Availability (count) of precinct amenities and facilities: Different facilities and amenities are implemented within each precinct to support residents' day-to-day needs and provide opportunities for community bonding. The types of facilities and amenities in public housing were combed through and classified according to the following categories (Figure 3.3):

Community and Precinct Facilities — Precinct Pavilions/ Amphitheatres, BBQ Pits, Open Plazas, Small Pavilions, Community Centres

Social Amenities — Kindergarten/ Childcare Services, Eldercare Services, Social Services/ Other VWOs, Healthcare Services, Residents' Committee

Sports and Recreational Facilities — Fitness Corners, Sports Courts, Playgrounds, Parks/ Common Green, Community Gardens

Commercial Amenities — Provision Kiosks, Coffeeshops/ Eating House, Hawker Centres, Neighbourhood Shops

Programmed Void Deck Space — Senior Citizens'/ Residents' Corner, Study Corner, Parklet

Precinct Centre — Precinct Centre

**FIGURE 3.3**   Precinct amenities and facilities in public housing in Singapore.

*Source:* Authors.

- Community and precinct facilities
- Social amenities
- Sports and recreational facilities
- Commercial amenities
- Programmed void deck space
- Precinct centre

3   Environmental performance: Factors like compactness, building proportions, and green spaces allocation significantly impact daylight performance and residents' overall QoL as they spend time in the precinct (Kim, 2017). The indicators for measuring environmental performance are:

- Aspect ratio (building height to building spacing, averaged at the precinct level)
- Site coverage ratio (area of building structures within a precinct to total surface area of a precinct)
- Area to perimeter ratio, also known as APR (building block area to block perimeter length, averaged at the precinct scale)

4 Accessibility: Accessibility variables quantify the number of facilities and amenities within a five-minute walking distance—approximately a 400 m radius, a rule òf thumb for a walkable catchment area (Daniels & Mulley, 2013)—from each residential block. These variables were collected by cross-referencing OneMap data (OneMap, 2018), the authoritative national map of Singapore, with onsite verification. The spatial data was mapped out on QGIS and the average number was calculated at the precinct level for the accessibility to the following:

- Community and precinct facilities
- Social amenities
- Sports and recreational facilities
- Commercial amenities

5 Pedestrian network: The pedestrian network primarily emphasised elements that contribute to pedestrian-friendly street design. This was assessed by the average number of signalised pedestrian crossings and the total length of covered linkways (Lane et al., 2020), calculated within a 400 m radius from each residential block and aggregated at the precinct level.

6 Inter-neighbourhood connectivity: Urban mobility facilitates access to different places and it is linked with a sense of independence, connectedness with the community, and a better QoL in general (Cooper et al., 2019). The indicators used to measure inter-neighbourhood connectivity are:

- Total number of bus stops within a 400 m radius from each residential block, averaged at the precinct level
- Distances to the nearest Mass Rapid Transit (MRT) or Light Rail Transit (LRT) stations from each residential block, averaged at the precinct level

### *Principal Component Analysis (PCA) and Cluster Analysis*

To advance to the second quantitative workflow, one of the MDC methods—PCA—was applied to determine the underlying factors responsible for classifying precinct typologies. The analysis included all 24 planning-related variables mentioned earlier and involved the generation of a cumulative scree plot with a cut-off explained variance (Figure 3.4). This plot enabled the identification of the first few principal components that were instrumental in the analysis, as they could collectively account for about 90% of the variance in the data. The loadings of each planning-related variable on the principal components were plotted with the

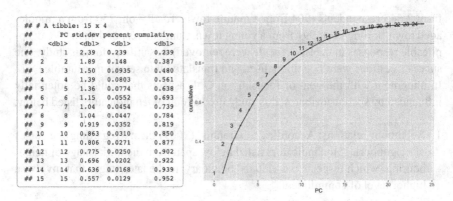

```
## # A tibble: 15 x 4
##       PC std.dev percent cumulative
##    <dbl>   <dbl>   <dbl>      <dbl>
## 1     1    2.39   0.239      0.239
## 2     2    1.89   0.148      0.387
## 3     3    1.50   0.0935     0.480
## 4     4    1.39   0.0803     0.561
## 5     5    1.36   0.0774     0.638
## 6     6    1.15   0.0552     0.693
## 7     7    1.04   0.0454     0.739
## 8     8    1.04   0.0447     0.784
## 9     9   0.919   0.0352     0.819
## 10   10   0.863   0.0310     0.850
## 11   11   0.806   0.0271     0.877
## 12   12   0.775   0.0250     0.902
## 13   13   0.696   0.0202     0.922
## 14   14   0.636   0.0168     0.939
## 15   15   0.557   0.0129     0.952
```

**FIGURE 3.4**    An example of a summary information of PCA (left) and a cumulative scree plot with a cut-off explained variance (right).

*Source:* Authors.

Varimax rotation method to enable a better visualisation for understanding the re-lationships among factors. Next, an Agglomerative Hierarchical Cluster Analysis was run using the principal components identified in the PCA. The agglomerative hierarchical clustering works in a bottom-up manner such that each observation starts as a single cluster (leaf) and goes through an iterative process to merge with other observations based on their similarities, until only one single, huge cluster is left containing all the observations (Bouguettaya et al., 2015; Day & Edelsbrunner, 1984; Murtagh & Contreras, 2012). Finally, $k$-means (the number of $k$ as suggested by the structure of the agglomerative algorithm) was applied to form the clusters, assigning each precinct to a specific precinct typology based on its similarities with respect to the factors. The entire process was independently applied to the three case study sites.

### Understanding and Collecting Data Relating to Residents' Perceived QoL

While QoL is a complex, multi-dimensional concept, this work specifically con-centrated on perceived QoL indicators that can be potentially influenced by spatial attributes in the surrounding built environment. These indicators were meticu-lously selected to enhance the understanding of people-environment research and provide insights for planners and architects. To understand residents' perceived QoL, the survey data collected for the New Urban Kampung Research Programme was employed in this study.

The door-to-door interviewer-administered survey required public housing resi-dents to rate their responses on a 5-point Likert scale ranging from 1 to 5, with higher scores indicating greater level of agreement or satisfaction. Data collected from 2,645 respondents (904 from a mature town, Toa Payoh; 750 from a middle-aged town, Jurong East; 991 from a young town, Punggol; respondents with not applicable responses in the Likert scale were removed) was utilised for this study. The survey questions were operationalised and translated into sets of measurable QoL. In alignment with the goal of focusing on QoL aspects that could be potentially influenced by spatial attributes, this study used a total of ten indicators listed below.

1   Neighbour relations: According to Grannis' (2009) theory regarding the stages of neighbouring, the indicators listed below span from the initial stage of neighbouring, which respects individuals' boundary, to the later stages that involve a higher level of commitment.

   • Hi-bye neighbouring
   • Activity-driven neighbouring
   • Support from neighbours

2   Quality of leisure facilities: This indicator refers to residents' satisfaction with the quality of leisure facilities provided in the precinct.
3   Civic participation: This indicator captures residents' perceived efficacy in neighbourhood civic participation, measuring their perceived influences on neighbourhood-related decisions.
4   Community participation: This indicator measures the perception of residents engaged in neighbourhood life through their participation in various community activities organised in their neighbourhood.
5   Sense of belonging: This indicator assesses residents' sense of belonging, with a particular focus on their desire to remain in the precinct and their cognitive or emotional sentiments towards the neighbourhood.
6   Mobility and transport: Recognising that movement is essential for residents performing daily routines and activities, this indicator evaluates their satisfaction with accessibility, including public transport connectivity, inter-neighbourhood connectivity, walkability, and accessibility to amenities and services.
7   Physical and built environment: This indicator encompasses the landscaping and elements that shape residents' living environment, including aspects such as noise pollution, density, maintenance, safety and security.
8   Well-being and lifestyle: Going beyond the narrow terms of physical or mental health, this indicator measures residents' overall state of being, such as perceived stress, quality of lifestyle, happiness, and quality of neighbourhood lifestyle.

### Statistical Analysis: Kruskal-Wallis Test and Post Hoc Analysis

This step involved a two-step analysis in order to answer the fourth part of quantitative workflow. First, the Kruskal-Wallis (Kruskal & Wallis, 1952) test was applied to determine whether there are statistically significant differences in residents' QoL among the identified precinct typologies. Second, if the Kruskal-Wallis test indicated significant differences amongst the groups, post hoc tests were run to identify which specific groups within the precinct typologies exhibit differing QoL.

## Case Studies

### *Toa Payoh, the Mature Town*

Previously a notorious squatter district, the place was cleared out in 1962 and began redevelopment in 1964. The town development was eventually completed in 1970 as the first satellite town holistically planned and developed by HDB (Urban Redevelopment Authority, 1995). With a total land area of 5.56 km², the mature town currently accommodates 39,737 dwelling units (as of 2019) organised into 41 precincts, with a total of 121,850 urban residents (as of 2020). Out of the entire town's population, 21.03% are residents aged 65 and over, 66.90% aged between 15 and 65, and 12.07% aged below 15 (Department of Statistics Singapore, 2020).

Through PCA and cluster analysis, a total of five unique precinct typologies (Figure 3.5) have emerged from the 41 precincts in Toa Payoh. Figure 3.6 shows the direction plot of the top ten principal components that jointly explain about 90% variance in the data as identified from the PCA. This direction plot allowed the understanding of each component's characteristics, and subsequently was translated into another direction plot (Figure 3.7) to visualise the combination of precinct characteristics for each precinct type. Table 3.1 shows the descriptive statistics of each precinct type in Toa Payoh.

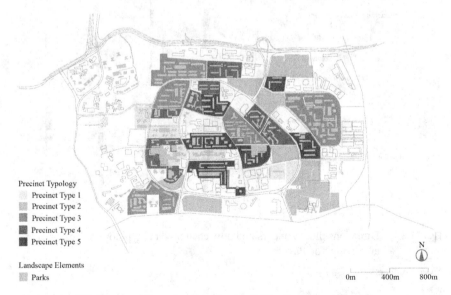

Precinct Typology
  Precinct Type 1
  Precinct Type 2
  Precinct Type 3
  Precinct Type 4
  Precinct Type 5

Landscape Elements
  Parks

N

0m    400m    800m

**FIGURE 3.5**   Distribution of precinct typologies in Toa Payoh.

*Source*: Authors.

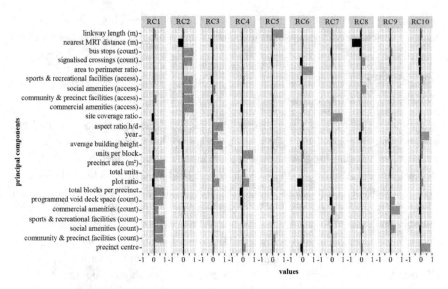

**FIGURE 3.6** Directions plot with the Varimax Method to understand the characteristics of each principal component (RC), explained by the loadings of each planning-related variable in Toa Payoh.

*Source*: Authors.

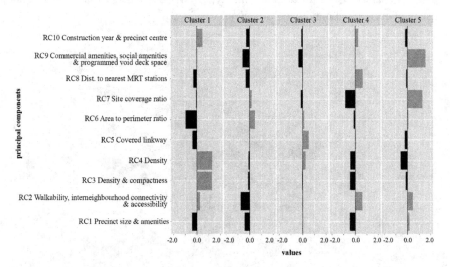

**FIGURE 3.7** Directions plot to understand how each precinct typology varies from one another in Toa Payoh.

*Source*: Authors.

TABLE 3.1 Descriptive statistics of each precinct typology in Toa Payoh showing variable means and standard deviation (in parentheses)

| | Precinct Type TPY1: Contemporary Solitary Skyscrapers | Precinct Type TPY2: Cosy Remote Enclaves | Precinct Type TPY3: Expansive Vintage Nexuses | Precinct Type TPY4: Edgewalk Estates | Precinct Type TPY5: Mid-sized Central Buzzes |
|---|---|---|---|---|---|
| Precinct density | | | | | |
| Plot ratio | 4.07 (0.10) | 3.00 (0) | 3.00 (0) | 3.00 (0) | 3.00 (0) |
| Total no. of units per precinct | 899 (289) | 493 (205) | 1890 (204) | 578 (211) | 771 (406) |
| Total no. of units per block | 228 (45.9) | 169 (96.2) | 156 (29.5) | 148 (44.6) | 104 (51.1) |
| Total no. of blocks per precinct | 3.50 (1.64) | 3.38 (2.06) | 12.0 (2.53) | 3.70 (1.64) | 8.17 (5.12) |
| Year of construction | 2004.83 | 1983.69 | 1973.83 | 1985.40 | 1973.83 |
| Precinct area (m$^2$) | 26,100 (4,940) | 26,300 (12,500) | 83,100 (18,100) | 28,400 (12,300) | 50,400 (17,200) |
| Average building ht. (m) | 90.1 (30.2) | 55.4 (24.8) | 61.2 (13.7) | 51.9 (17.0) | 41.5 (26.5) |
| Precinct amenities and facilities | | | | | |
| Community and precinct facilities | 1.50 (1.38) | 1.77 (2.09) | 7.67 (3.01) | 2.00 (1.41) | 2.83 (1.60) |
| Social amenities | 2.00 (2.53) | 1.46 (1.66) | 9.33 (3.83) | 3.30 (2.75) | 5.83 (2.64) |
| Sports and recreational facilities | 1.83 (1.83) | 3.31 (2.10) | 10.7 (2.94) | 3.10 (1.20) | 4.17 (1.17) |
| Commercial amenities | 0.33 (0.52) | 0.54 (0.97) | 3.00 (1.41) | 1.40 (1.96) | 6.83 (1.72) |
| Programmed void deck spaces | 0.83 (0.75) | 1.23 (1.24) | 6.33 (3.01) | 2.10 (1.20) | 3.83 (1.72) |
| Precinct centre | 0.67 | 0.08 | 0.33 | 0.30 | 0.00 |
| Environmental performance | | | | | |
| Aspect ratio | 3.87 (1.78) | 2.28 (1.31) | 2.71 (0.70) | 2.28 (1.19) | 2.52 (1.77) |
| Site coverage ratio | 0.23 (0.08) | 0.27 (0.08) | 0.19 (0.10) | 0.19 (0.07) | 0.44 (0.15) |

(Continued)

TABLE 3.1 (Continued)

| | Precinct Type TPY1: Contemporary Solitary Skyscrapers | Precinct Type TPY2: Cosy Remote Enclaves | Precinct Type TPY3: Expansive Vintage Nexuses | Precinct Type TPY4: Edgewalk Estates | Precinct Type TPY5: Mid-sized Central Buzzes |
|---|---|---|---|---|---|
| Area to perimeter ratio | 4.96 (2.05) | 6.40 (0.70) | 6.30 (0.61) | 5.54 (0.83) | 6.15 (0.33) |
| Accessibility | | | | | |
| Community and precinct facilities | 18.7 (2.95) | 15.4 (5.46) | 23.4 (7.84) | 23.6 (5.59) | 23.3 (5.12) |
| Social amenities | 32.3 (11.7) | 19.3 (8.07) | 30.1 (9.03) | 37.3 (8.63) | 37.6 (11.4) |
| Sports and recreational facilities | 30.3 (3.96) | 23.2 (6.60) | 29.8 (7.24) | 32.5 (4.17) | 29.3 (6.72) |
| Commercial amenities | 13.7 (4.14) | 9.68 (4.68) | 14.7 (6.95) | 17.9 (7.36) | 18.4 (5.19) |
| Pedestrian network | | | | | |
| Signalised crossings | 18.5 (3.12) | 14.0 (5.62) | 16.5 (6.45) | 20.2 (3.55) | 19.2 (4.18) |
| Total length of covered linkway (m) | 75.5 (50.7) | 85.7 (37.3) | 128 (40.0) | 95.2 (51.8) | 83.8 (53.2) |
| Inter-neighbourhood connectivity | | | | | |
| Total no. of bus stops | 12.0 (2.46) | 8.67 (3.26) | 10.8 (2.19) | 12.3 (2.43) | 12.8 (2.03) |
| Distance to the nearest MRT/LRT stations (m) | 508 (253) | 905 (264) | 655 (469) | 430 (222) | 568 (334) |

*Precinct Typology TPY1 (n = 6): Contemporary Solitary Skyscrapers*

As the newest (averagely built in 2004.83) and smallest group of precincts (26,100.00 m²), the *Contemporary Solitary Skyscrapers*, situated in Toa Payoh's inner ring, are characterised by their compact yet tall residential blocks, boasting the highest average building height (90.1 m) and unit number (228.0). Their limited precinct sizes naturally result in the lowest average number in most of the amenities within the precinct boundaries, leading to moderate accessibility to different amenities within a 400 m radius of each block. While having a fair average number of signalised crossings, these precincts have the shortest average length (75.5 m) of covered linkway compared to other precinct types. As for inter-neighbourhood connectivity, residents may have more leeway to choose between buses or MRT for their daily commute, given the reasonable number of bus stops scattered around and the average walkable distance (508.0 m) to the nearest MRT/LRT station.

*Precinct Typology TPY2 (n = 13): Cosy Remote Enclaves*

Located mainly on the northeast and southeast perimeters of Toa Payoh, the *Cosy Remote Enclaves* are the second-smallest precincts (averaged at 26,300.00 m²). Given their compact precinct sizes and remote locations, they offer a limited number of facilities and amenities, especially social amenities, within a precinct. These precincts also record the lowest accessibility to all types of facilities and amenities, as well as to signalised crossings and bus stops. Additionally, their geographic location results in a substantial distance (905.0 m, the furthest) to the nearest MRT/LRT station, with a moderate provision of covered linkway (85.7 m).

*Precinct Typology TPY3 (n = 6): Expansive Vintage Nexuses*

As one of the oldest precinct groups, constructed in the early 1970s, the *Expansive Vintage Nexuses* stand out with a unique Y-shaped block (in one of the precincts) and the largest average precinct size (83,100.00 m²) with high-rise, high-density residential blocks. Located mainly on Toa Payoh's outskirts, these precincts are characterised with a diverse mix of amenities, particularly precinct and community facilities, social amenities, sports and recreational facilities, and programmed void deck spaces. This results in good accessibility to different types of amenities, accompanied by the longest average length (128.0 m) of covered linkways within a precinct boundary, yet a relatively low average number (16.5) of signalised crossings. This precinct type has a moderate inter-neighbourhood connectivity, given a fair number of bus stops scattered around and an acceptable walking distance (averaged at 655.0 m) to the nearest MRT/LRT station.

*Precinct Typology TPY4 (n = 10): Edgewalk Estates*

Mainly built during the mid-1980s, the *Edgewalk Estates* are distinguished by their high-rise, high-density residential blocks. Situated along the periphery of

Toa Payoh's inner ring, these precincts offer excellent accessibility to community precinct facilities, as well as sports and recreational facilities. Their advantageous location results in very good inter-neighbourhood connectivity, with the shortest average distance (430.0 m) to the nearest MRT/LRT station and the second-highest average number (12.3) of bus stops. This precinct type also boasts a good pedestrian network, featuring the highest average number (20.2) of signalised crossings and a generously long (95.2 m) covered linkway.

### Precinct Typology TPY5 (n = 6): Mid-sized Central Buzzes

The *Mid-sized Central Buzzes*, built in the early 1970s, stand as one of the oldest precinct groups. Notable for their commercial amenities such as neighbourhood shops and eateries, these precincts have the lowest average building height (41.48 m) among all types. With the second-largest average precinct size (50,400.00 m²), they provide a considerable number of residential blocks with interspersed amenities. Positioned centrally, this precinct type offers excellent accessibility to different types of amenities, particularly commercial and social amenities. For inter-neighbourhood connectivity, residents may rely on buses for their daily commutes due to the abundance of bus stops (12.8, highest) scattered around. The average distance (568.0 m) to the nearest MRT/LRT station is relatively on par. While the average length (83.8 m) of covered linkway is relatively low, residents living in this precinct type enjoy a safe and walkable environment, given the generous number (19.2, second-highest) of signalised crossings.

### Precinct Typologies and Residents' Perceived QoL

The Kruskal-Wallis test was conducted to assess statistically significant differences in residents' perceived QoL among the five identified precinct typologies in Toa Payoh. However, the results indicated that no significant differences were observed.

### Jurong East, the Middle-aged Town

In the 1960s, Jurong underwent development to transform the former swampland into Singapore's first industrial estate. Only in the 1970s and the 1990s did the construction of housing estates and amenities gain momentum, aiming to create self-sufficient new towns and attracting more workers and companies to the town. This transformation subsequently urged the massive relocation of people, many of whom were from the same villages, to this new town (Cho et al., 2018; Housing & Development Board, 2020a). Located at Singapore's southwestern tip with a total land area of 3.84 km², Jurong East today is home to 24,122 residential flats (as of 2019) arranged into 26 precincts, with a total town population of 78,600 (as of 2020). The town population is made up of 19.03% of residents aged 65 and

**FIGURE 3.8**    Distribution of precinct typologies in Jurong East.

*Source*: Authors.

above, 69.60% aged between 15 and 65, and 11.37% aged below 15 (Department of Statistics Singapore, 2020).

Within the 26 precincts, the use of PCA and cluster analysis has revealed four unique precinct types in Jurong East (Figure 3.8). The direction plot in Figure 3.9 depicts the first nine principal components, collectively explaining approximately 90% of the variance in the data. These principal components were further condensed and represented in the direction plot in Figure 3.10, capturing a combination of precinct characteristics used to characterise the four identified precinct types. Refer to Table 3.2 for the descriptive statistics of each precinct type in Jurong East.

### Precinct Typology JE1 (n = 10): Compact Secluded Nooks

Built in the late 1980s, the *Compact Secluded Nooks* occupy Jurong East's northwestern tip, distinguished by their smallest average precinct size (36,300.00 m²) and fewest blocks per precinct (4.7). Despite these constraints, they boast the highest average number of units per block (119.0), attributed to their highest average building height (45.2 m) among the four precinct types. Although social

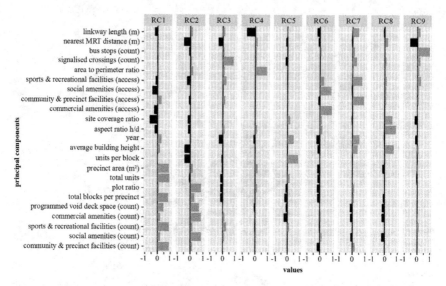

**FIGURE 3.9**  Directions plot with the Varimax Method to understand the characteristics of each principal component (RC), explained by the loadings of each planning-related variable in Jurong East.

*Source*: Authors.

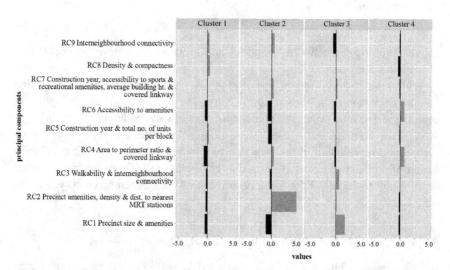

**FIGURE 3.10**  Directions plot to understand how each precinct typology varies from one another in Jurong East.

*Source*: Authors.

**TABLE 3.2** Descriptive statistics of each precinct typology in Jurong East showing variable means and standard deviation (in parentheses)

| | Precinct Type JE1: Compact Secluded Nooks | Precinct Type JE2: Low-Rise Railside Plaza | Precinct Type JE3: Outskirt Supreme Residences | Precinct Type JE4: Mid-Rise Serene Hubs |
|---|---|---|---|---|
| Precinct density | | | | |
| Plot ratio | 2.80 (0) | 3.50 (NA) | 2.87 (0.16) | 2.80 (0) |
| Total no. of units per precinct | 546 (258) | 70.0 (NA) | 1,350 (607) | 537 (193) |
| Total no. of units per block | 119 (44.1) | 10.0 (NA) | 101 (9.03) | 107 (37.0) |
| Total no. of blocks per precinct | 4.70 (2.06) | 11.0 (NA) | 13.0 (5.00) | 5.60 (3.03) |
| Year of construction | 1988.30 | 1986.00 | 1992.80 | 1983.80 |
| Precinct area (m$^2$) | 36,300 (24,300) | 59,200 (NA) | 103,000 (47,200) | 43,800 (13,000) |
| Average building ht. (m) | 45.2 (15.5) | 8.94 (NA) | 43.6 (12.1) | 30.6 (7.10) |
| Precinct amenities and facilities | | | | |
| Community and precinct facilities | 2.40 (2.17) | 4.00 (NA) | 11.8 (5.45) | 1.60 (1.43) |
| Social amenities | 2.10 (1.73) | 16.0 (NA) | 5.80 (3.56) | 4.30 (3.27) |
| Sports and recreational facilities | 3.50 (3.47) | 4.00 (NA) | 10.0 (1.41) | 3.00 (2.11) |
| Commercial amenities | 1.10 (0.88) | 12.0 (NA) | 4.40 (1.82) | 2.90 (2.56) |
| Programmed void deck spaces | 2.30 (0.95) | 10.0 (NA) | 6.20 (3.03) | 3.50 (1.27) |
| Precinct centre | 0.00 | 0.00 | 0.00 | 0.00 |
| Environmental performance | | | | |
| Aspect ratio | 1.19 (1.29) | 0.14 (NA) | 0.41 (0.14) | 0.88 (0.66) |
| Site coverage ratio | 0.05 (0.02) | 0.02 (NA) | 0.02 (0.01) | 0.04 (0.02) |

(Continued)

**TABLE 3.2** (Continued)

| | Precinct Type JE1: Compact Secluded Nooks | Precinct Type JE2: Low-Rise Railside Plaza | Precinct Type JE3: Outskirt Supreme Residences | Precinct Type JE4: Mid-Rise Serene Hubs |
|---|---|---|---|---|
| Area to perimeter ratio | 5.46 (0.95) | 7.11 (NA) | 5.81 (0.90) | 6.51 (0.47) |
| Accessibility | | | | |
| Community and precinct facilities | 20.7 (7.42) | 17.6 (NA) | 25.7 (6.73) | 17.8 (6.97) |
| Social amenities | 25.9 (7.25) | 28.2 (NA) | 18.1 (8.57) | 32.1 (5.31) |
| Sports and recreational facilities | 29.7 (7.59) | 22.3 (NA) | 27.0 (3.82) | 29.5 (6.78) |
| Commercial amenities | 14.7 (3.87) | 16.3 (NA) | 12.2 (5.08) | 21.5 (5.65) |
| Pedestrian network | | | | |
| Signalised crossings | 12.6 (3.45) | 16.2 (NA) | 15.3 (6.69) | 13.4 (2.89) |
| Total length of covered linkway (m) | 47.4 (14.6) | 58.0 (NA) | 36.0 (22.4) | 22.0 (14.1) |
| Inter-neighbourhood connectivity | | | | |
| Total no. of bus stops | 9.06 (1.90) | 8.27 (NA) | 7.95 (2.03) | 8.63 (2.26) |
| Distance to the nearest MRT/LRT stations (m) | 711 (166) | 313 (NA) | 696 (176) | 729 (153) |

*Source:* Authors.

amenities, programmed void deck spaces, and commercial amenities within these precincts are limited, they maintain fair accessibility, especially to sports and recreational facilities. These precincts also have the highest aspect ratio and site coverage ratio, indicating potential shadow casting and limited open green space. While positioned farther (711.0 m, second-furthest) from the nearest MRT/LRT station compared to other groups, these precincts have the highest number (9.06) of bus stops scattered around. The residents also benefit from a network of covered linkways (second-highest), albeit with the lowest average number of signalised crossings (12.6).

### Precinct Typology JE2 (n = 1): Low-rise Railside Plaza

The *Low-Rise Railside Plaza* features a single, distinctive precinct known for its low-density residential blocks, spaciousness (average precinct size of 59,200.00 m², second-largest), and the lowest average building height (8.94 m) among the four precinct types. Its precinct size allows for different types of precinct amenities and facilities, especially social and commercial amenities and programmed void deck spaces. However, situated on Jurong East's northwestern edge, certain residential blocks may not enjoy the same level of accessibility as others, as amenities and facilities are prone to be saturated in specific areas within the precinct. Notably, this precinct boasts the most extensive pedestrian network within the town, marked by the highest average number (16.2) of signalised crossings and the greatest average length of covered linkways. Despite having a relatively low number (8.27) of bus stops, this precinct has the shortest average distance to the nearest MRT/LRT station (313.0 m).

### Precinct Typology JE3 (n = 5): Outskirt Supreme Residences

Constructed in the early 1990s, the *Outskirt Supreme Residences* distinguish themselves with an extensive average precinct size (103,000.00 m²), accommodating the highest average number of residential blocks (13.0) and units (1,350.0). This spaciousness allows for a wide range of community and precinct facilities, as well as ample sports and recreational options, with good accessibility, though it lags in terms of social and commercial amenities. Their pedestrian network design is moderate, given a good average number (15.3) of signalised crossings and a fair length (36.0 m) of covered linkways. For inter-neighbourhood connectivity, these precincts have the lowest average number (7.95) of bus stops, but the average distance (696.0 m) to the nearest MRT/LRT stations is on par with other precinct types.

### Precinct Typology JE4 (n = 10): Mid-rise Serene Hubs

The *Mid-rise Serene Hubs* are made up of the oldest (averagely built in 1,983.80) group of precincts in Jurong East. Characterised by mid-rise residential blocks (average height 30.6 m), they house the lowest average number of units per residential block (107.0) in the middle-aged town. With a moderate average precinct size (43,800.00 m²), these precincts offer very few amenities and facilities within their boundaries, but boast the highest accessibility to social and commercial amenities, mostly from their neighbouring precincts. However, their pedestrian network falls short, with a low average number (13.4) of signalised crossings and the shortest average length (22.0 m) of covered linkways. Inter-neighbourhood connectivity is subpar, marked by a fair average number (8.63) of bus stops and the longest average distance (729.0 m) to the nearest MRT/LRT station.

### Precinct Typologies and Residents' Perceived QoL

The Kruskal-Wallis test was used to evaluate statistically significant differences in residents' perceived QoL within the four identified precinct typologies in Jurong East. Nevertheless, similar to the case observed in Toa Payoh, the findings revealed no significant difference in residents' perceived QoL despite the different precinct typologies they live in.

### Punggol, the Young Town

Situated in Singapore's northeast region, Punggol was formerly home to settlers who engaged in fishing, plantation work, poultry farming, and pig-rearing activities. Punggol of today, however, is very different from its rural past. This transformation began in 1996 with the announcement of redevelopment and reclamation work, aiming to transform the old village into a waterfront town with housing along the river and coastal area. Significant emphasis has been placed in portraying the young town as Singapore's first 'eco-town', as well as a 'living laboratory' for HDB to test various novel ideas and technologies in sustainable development (Housing & Development Board, 2020b). With a total land area of 9.57 km², the young town currently houses 50,663 dwelling units (as of 2019) organised into 66 precincts, accommodating 174,450 residents (as of 2020). Among the residents, 7.11% are aged 65 and over, 68.3% are aged between 15 and 65, and 24.59% are those aged below 15 (Department of Statistics Singapore, 2020).

A total of six distinctive precinct typologies (Figure 3.11) have been discovered by applying PCA and cluster analysis to the 66 precincts in Punggol. Figure 3.12 presents the direction plot of the first 12 principal components, collectively explaining approximately 90% of the variance in the data. This plot provided insights into the characteristics of each component. These insights were further summarised and represented in another direction plot (Figure 3.13) to visualise the combination of precinct characteristics for describing each of the identified precinct types. Table 3.3 shows the descriptive statistics of each precinct type in Punggol.

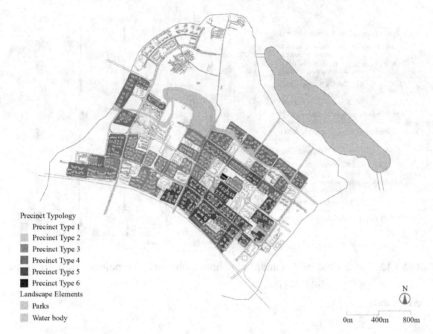

**FIGURE 3.11**    Distribution of precinct typologies in Punggol.

*Source*: Authors.

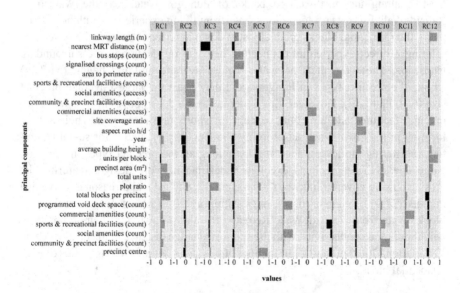

**FIGURE 3.12**    Directions plot with the Varimax Method to understand the characteristics of each principal component (RC), explained by the loadings of each planning-related variable in Punggol.

*Source*: Authors.

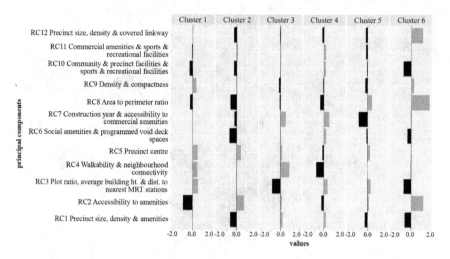

**FIGURE 3.13** Directions plot to understand how each precinct typology varies from one another in Punggol.

*Source*: Authors.

## Precinct Typology PG1 (n = 2): Waterfront RailsideWalk Retreats

Nestled along the northwestern banks of Punggol Waterway, the Waterfront RailsideWalk Retreats offer a unique waterfront living experience. With hexagonal plans, the residential blocks gradually step down in terraces towards the waterway, offering a direct experiential connection between residents and the surrounding landscape. These precincts have the highest average building height (51.2 m) among the precinct types in Punggol, and they boast the largest average precinct size (65,000.00 m²). However, this generous space does not result in a wealth of precinct facilities and amenities, except for sports and recreation. Their remote location in the town poses accessibility challenges, particularly for social amenities, registering the lowest accessibility among all the precinct types in Punggol. Nonetheless, these precincts excel in pedestrian network design, featuring the most extensive covered linkways (averaged at 195.0 m) and a reasonable average number (21.0) of signalised crossings within a 400 m radius of each residential block. Waterfront RailsideWalk Retreats are also known for their good inter-neighbourhood connectivity, with the shortest average distance to the nearest MRT/LRT station (200.0 m) and a fair number of bus stops within a 400 m radius of each residential block.

## Precinct Typology PG2 (n = 2): Modest ModernScape Heights

The *Modest ModernScape Heights* boast the second-smallest (12,800.00 m²), newest groups of precincts in Punggol, with an average construction year of

**TABLE 3.3** Descriptive statistics of each precinct typology in Punggol showing variable means and standard deviation (in parentheses)

| | Precinct Type PG1: Waterfront RailsideWalk Retreats | Precinct Type PG2: ModernScape Heights | Precinct Type PG3: Central Mercantile Residences | Precinct Type PG4: Countryside Marketplace Vistas | Precinct Type PG5: Timeless Connected Charms | Precinct Type PG6: MicroLink Living |
|---|---|---|---|---|---|---|
| Precinct density | | | | | | |
| Plot ratio | 3.20 (0) | 3.30 (0.14) | 3.08 (0.16) | 3.24 (0.20) | 3.27 (0.22) | 3.00 (NA) |
| Total no. of units per precinct | 938 (190) | 327 (139) | 792 (344) | 881 (251) | 622 (169) | 512 (NA) |
| Total no. of units per block | 313 (62.9) | 92.5 (30.4) | 107 (29.4) | 115 (29.6) | 98.5 (21.6) | 256 (NA) |
| Total no. of blocks per precinct | 3.00 (0) | 4.00 (2.83) | 7.50 (2.46) | 8.10 (3.06) | 6.45 (1.71) | 2.00 (NA) |
| Year of construction | 2013.00 | 2016.00 | 2012.78 | 2015.38 | 2004.05 | 2010.00 |
| Precinct area (m²) | 65,000 (25,400) | 12,800 (1,600) | 33,100 (10,700) | 42,200 (14,400) | 29,100 (8,040) | 7,330 (NA) |
| Average building ht. (m) | 51.2 (0.66) | 50.3 (1.98) | 41.2 (1.74) | 45.4 (3.44) | 45.0 (3.61) | 48.9 (NA) |
| Precinct amenities and facilities | | | | | | |
| Community and precinct facilities | 1.50 (0.71) | 0.50 (0.71) | 2.78 (2.32) | 3.62 (2.13) | 2.36 (1.62) | 0 (NA) |
| Social amenities | 1.50 (2.12) | 0 (0) | 1.50 (1.25) | 2.05 (1.56) | 1.27 (1.16) | 1.00 (NA) |
| Sports and recreational facilities | 4.50 (2.12) | 1.00 (1.41) | 4.44 (2.25) | 4.52 (2.29) | 2.64 (1.56) | 0 (NA) |
| Commercial amenities | 0 (0) | 0 (0) | 0.67 (0.91) | 0.95 (1.07) | 0.18 (0.50) | 0 (NA) |
| Programmed void deck spaces | 0.50 (0.71) | 0 (0) | 2.00 (1.33) | 2.48 (1.72) | 1.68 (1.39) | 0 (NA) |
| Precinct centre | 1.00 | 1.00 | 1.00 | 1.00 | 1.00 | 0 |

(Continued)

**TABLE 3.3** (Continued)

| | Precinct Type PG1: Waterfront RailsideWalk Retreats | Precinct Type PG2: ModernScape Heights | Precinct Type PG3: Central Mercantile Residences | Precinct Type PG4: Countryside Marketplace Vistas | Precinct Type PG5: Timeless Connected Charms | Precinct Type PG6: MicroLink Living |
|---|---|---|---|---|---|---|
| Environmental performance | | | | | | |
| Aspect ratio | 1.15 (0.04) | 7.96 (3.08) | 0.84 (0.18) | 0.81 (0.34) | 1.13 (0.51) | 1.53 (NA) |
| Site coverage ratio | 0.11 (0.07) | 0.10 (0.02) | 0.03 (0.01) | 0.02 (0.01) | 0.04 (0.02) | 0.12 (NA) |
| Area to perimeter ratio | 3.32 (0.28) | 3.05 (0.39) | 3.17 (0.34) | 3.05 (0.20) | 3.53 (0.44) | 4.71 (NA) |
| Accessibility | | | | | | |
| Community and precinct facilities | 17.0 (0) | 26.0 (12.0) | 18.8 (5.53) | 20.0 (5.92) | 22.3 (6.03) | 30.5 (NA) |
| Social amenities | 11.8 (1.06) | 25.0 (14.1) | 23.8 (5.02) | 17.9 (7.30) | 21.3 (7.97) | 34.5 (NA) |
| Sports and recreational facilities | 21.0 (4.24) | 30.3 (9.55) | 30.9 (5.85) | 24.0 (6.13) | 28.8 (5.99) | 32.5 (NA) |
| Commercial amenities | 3.00 (0) | 2.50 (3.54) | 5.17 (1.42) | 4.37 (1.50) | 2.70 (1.62) | 3.00 (NA) |
| Pedestrian network | | | | | | |
| Signalised crossings | 21.0 (4.24) | 21.5 (2.83) | 25.7 (5.23) | 11.9 (6.31) | 18.5 (4.80) | 16.0 (NA) |
| Total length of covered linkway (m) | 195 (61.6) | 36.5 (6.01) | 45.1 (41.7) | 13.3 (23.1) | 38.8 (13.2) | 45.6 (NA) |
| Inter-neighbourhood connectivity | | | | | | |
| Total no. of bus stops | 7.75 (0.35) | 8.25 (1.77) | 7.78 (2.31) | 4.39 (2.56) | 6.80 (1.72) | 9.00 (NA) |
| Distance to the nearest MRT/LRT stations (m) | 200 (62.6) | 226 (94.7) | 357 (143) | 262 (161) | 203 (64.9) | 332 (NA) |

*Source*: Authors.

2016. These high-rise developments, which have the second-highest average building height (50.3 m) in Punggol, interestingly have the lowest average number of units per precinct (327.0) and per block (92.5). Similar to *Waterfront RailsideWalk Retreats*, this precinct type is relatively ill-equipped, lacking social amenities, programmed void deck space, and commercial amenities. However, this is offset by reasonable accessibility to different precinct facilities and amenities, especially sports and recreational facilities, social amenities, and community and precinct facilities, due to its proximity to neighbouring precincts. The pedestrian network for this precinct type is moderately planned, featuring a good average number (21.5) of signalised crossings but with a very limited average length (36.5 m) of covered linkways. Nevertheless, these precincts excel in inter-neighbourhood connectivity, offering a reasonable average number (8.25) of bus stops and a convenient walking distance to the nearest MRT/LRT station, averaging 226.0 m.

### Precinct Typology PG3 (n = 18): Central Mercantile Residences

Located mainly in the heart of Punggol, the *Central Mercantile Residences* comprise precincts with the town's lowest average building height (41.16 m). These precincts offer a reasonable number of different facilities and amenities within a precinct boundary, especially for sports and recreational facilities, and boast the best accessibility to commercial amenities among all precinct types in Punggol. While having the highest average number (25.7) of signalised crossings within a 400 m radius of each residential block, the covered linkways are relatively short (45.1 m). Though these precincts are the furthest from the MRT/LRT station (average distance 357.0 m), they have a manageable walking distance, with a moderate number of bus stops (7.78) within a 400 m radius of each residential block.

### Precinct Typology PG4 (n = 21): Countryside Marketplace Vistas

Nestled in the town's outskirts, residents from about half of this precinct type enjoy scenic waterfront and park views. The *Countryside Marketplace Vistas* are known for their best-equipped precincts in town, featuring the highest number of all types of facilities and amenities within a precinct boundary. Despite its remote location, these precincts offer moderate accessibility to most facilities, with exceptional access to commercial amenities, second only to *Central Mercantile Residences*. However, this precinct type lacks in its pedestrian network design, featuring the lowest average number of signalised crossings. It also has the shortest average length of covered linkways at just 13.3 m, significantly compromising the walkability within the town. In terms of inter-neighbourhood connectivity, considering the limited number (4.39) of bus stops within a convenient walking distance, residents from this precinct type may rely more on train commute, with an average distance of 262.0 m to the nearest MRT/LRT station.

## Precinct Typology PG5 (n = 22): Timeless Connected Charms

Located in the southeast corner of Punggol, the *Timeless Connected Charms* are made up of the town's oldest precincts, with an average construction year dating back to 2004. These precincts offer a reasonable selection of facilities and amenities within their precinct boundaries. Accessibility to various facilities and amenities are generally good, except for commercial amenities. While *Timeless Connected Charms* offer a fair number of signalised crossings within a manageable walking distance from each residential block, the average length of covered linkways is relatively short (38.8 m). However, this precinct type boasts a good inter-neighbourhood connectivity, featuring the second-shortest average distance to the nearest MRT/LRT station (203.0 m), and a fair number of bus stops within a 400 m radius of each residential block.

## Precinct Typology PG6 (n = 1): MicroLink Living

The *MicroLink Living* stands out as a distinct precinct type in Punggol: it is the only type without a precinct centre. *MicroLink Living* has the smallest precinct size (7,330.00 m$^2$), accommodating only a single social amenity within its precinct boundary. This is, however, offset by its highest accessibility to neighbouring precinct facilities and amenities within a 400 m radius, except for commercial amenities. This precinct lacks in pedestrian network design, as it has the second-lowest average number (16.0) of signalised crossings among all precinct types in Punggol and a moderate length (45.6 m) of covered linkways within the precinct. Nevertheless, it is compensated by its good inter-neighbourhood connectivity, boasting the highest average number (9.0) of bus stops and a convenient walking distance (average at 332.0 m) to the nearest MRT/LRT station.

## Precinct Typologies and Residents' Perceived QoL

To assess residents' perceived QoL in Punggol, the Kruskal-Wallis test along with post hoc analysis, specifically the multiple comparisons Dunn's test, were employed. This approach allowed the identification of statistically significant differences in QoL among residents residing in different precinct typologies in Punggol. Notably, unlike the cases observed in Toa Payoh and Jurong East, the analysis revealed significant differences in residents' perceived QoL between specific pairs of precinct types in Punggol (Table 3.4 and Figure 3.14).

Particularly, residents living in *Timeless Connected Charms (PG5)* rated significantly higher sense of belonging ($p < .05$) and higher satisfaction with the physical and built environment ($p < .05$) than those living in *Countryside Marketplace Vistas (PG4)*.

The indicator for sense of belonging was measured with a list of statements related to residents' emotional attachment to their physical living environment over time, including feeling of belonging to the neighbourhood; intentions to remain

**TABLE 3.4**  Post hoc analysis (multiple comparisons Dunn's test) in Punggol

| Indicators | Comparison of Precinct Types | Mean and Standard Deviation | | Chi-Square | Significance Level (p) |
|---|---|---|---|---|---|
| Sense of belonging | PG5 Timeless Connected Charms and PG4 Countryside Marketplace Vistas | PG5 3.75 (0.21) | PG4 3.55 (0.21) | 13.9168 | 0.0493* |
| Physical and built environment | PG5 Timeless Connected Charms and PG4 Countryside Marketplace Vistas | PG5 3.22 (0.26) | PG4 3.10 (0.16) | 13.9352 | 0.0155* |
| Support from neighbours | PG1 Waterfront RailsideWalk Retreats and PG2 ModernScape Heights | PG1 3.66 (0.30) | PG2 2.83 (0.24) | 11.9214 | 0.0467* |
| Mobility and transport | PG3 Central Mercantile Residences & PG4 Countryside Marketplace Vistas | PG3 3.29 (0.32) | PG4 3.05 (0.28) | 11.8697 | 0.0482* |

Source: Authors.

$*p < .05; **p < .01; ***p < .001.$

in residency in the neighbourhood for the next five years; satisfaction with the neighbourhood's physical beauty, etc. *Timeless Connected Charms (PG5)* were developed earlier than all the other precincts in Punggol (average construction year of 2004), in contrast to *Countryside Marketplace Vistas (PG4),* the newest precincts (average construction year of 2015). This suggests that residents with a longer duration of stay may contribute significantly to a higher sense of belonging, supporting the suggestions by Hidalgo and Hernandez (2001) and Kamalipour et al. (2012) that individuals' attachments to their neighbourhood grow with their temporal investments, which is often attributed to community involvement and the establishment of robust social networks (Brown et al., 2005; Filkins et al., 2000; Fleury-Bahi et al., 2008). These factors, in turn, determine their commitment to remain in residency in the neighbourhood (Grzeskowiak et al., 2003).

In relation to the physical and built environment, the QoL indicator considered residents' satisfaction with noise levels, residential density, changes of neighbourhood over time, and safety. Referring to Figure 3.14, *Timeless Connected Charms*

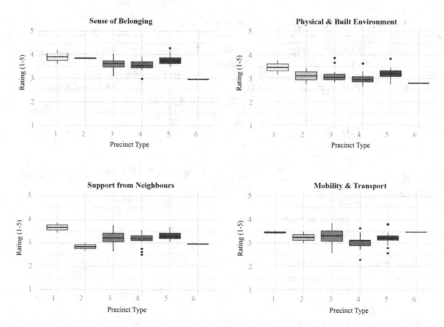

**FIGURE 3.14** Value distribution for QoL indicator of each precinct type in Punggol.
*Source*: Authors.

*(PG5)* are observed to have a higher residential density compared to *Country-side Marketplace Vistas (PG4)*. Specifically, in RC1 precinct size, density and amenities, *Timeless Connected Charms (PG5)* display a negative loading while *Countryside Marketplace Vistas (PG4)* exhibit a positive loading. Despite having a higher residential density, which is often associated with noise pollution as suggested by Chan and Lee (2009), residents of *Timeless Connected Charms (PG5)* reported higher satisfaction with the physical and built environment in their precincts. This could be attributed to the fact that they have witnessed and experienced the entire transformative town development made over the past two decades. This transformation included turning the town into an 'eco-town' with a 'waterfront' and high-rise living shifting away from the traditional public housing design (Housing & Development Board, 2020b), as initiated after *Timeless Connected Charms (PG5)* were established and potentially before *Countryside Marketplace Vistas (PG4)* were developed. Residents of *Countryside Marketplace Vistas (PG4)* might also contend with noise pollution from Seletar Airport, situated just 6.0 km east of Punggol, where many precincts from *Countryside Marketplace Vistas (PG4)* are located.

Furthermore, the post hoc analysis revealed that residents from *Waterfront RailsideWalk Retreats (PG1)* rated significantly higher in their perception of receiving support from neighbours than those in *ModernScape Heights (PG2)*

($p < .05$). The QoL indicator was measured by residents' extents of being able to exchange favour with their neighbours, having mutual trust with their neighbours by depending on them in times of emergency, and sharing the same values and norms with their neighbours. Both *Waterfront RailsideWalk Retreats (PG1)* and *ModernScape Heights (PG2)* are similar in terms of having limited facilities and amenities within their precinct boundaries, yet residents of *ModernScape Heights (PG1)* enjoy slightly better accessibility to facilities and amenities from their neighbouring precincts. What sets *Waterfront RailsideWalk Retreats (PG1)* apart from *ModernScape Heights (PG2)* is the significant contrast in their pedestrian network design, particularly in the average length of covered linkways—the former *(PG1)* boasts an impressive average length of 195.0 m while the latter *(PG2)* lags behind with an average of only 35.5 m. As discussed earlier, these covered linkways serve not only as convenient routes to nearby facilities but also as spaces that promote spontaneous social interactions and potentially deeper connections with neighbours during daily commutes. Further research is needed to delve into the impacts of covered linkway layouts on different stages of neighbourliness (Grannis, 2009). It is also worth noting that *Waterfront RailsideWalk Retreats (PG1)* feature a unique architectural design for their housing complex with hexagonal plans. The design incorporates a continuous, no-dead-end corridor that runs the length of every floor, but rearranged in a courtyard formation. This design was derived from the traditional slab block typology, which is known for its efficacy in promoting neighbourly interactions in the corridors (Menz, 2014). This indicates the necessity for future research to conduct a more fine-grained study, specifically examining the impacts of block morphologies on varying stages of neighbourliness.

Based on the post hoc analysis, residents living in *Central Mercantile Residences (PG3)* also reported significantly higher satisfaction with mobility and transport than those from *Countryside Marketplace Vistas (PG4)* ($p < .05$). The QoL indicator for mobility and transport considered residents' satisfaction with public transport services in the neighbourhood, connection to regularly frequented places, neighbourhood walkability, and accessibility to facilities and amenities. As illustrated in Table 3.4, *Central Mercantile Residences (PG3)* generally exhibit better accessibility to a wide range of facilities and amenities, especially for commercial amenities, sports and recreational facilities, and social amenities. Additionally, *Central Mercantile Residences (PG3)* stand out for their pedestrian network design, boasting the highest number of signalised crossings and a reasonably extensive covered linkway network. In contrast, *Countryside Marketplace Vistas (PG4)* rank lowest in both aspects in Punggol, highlighting its significantly compromised walkability. The abovementioned is also evident in Figure 3.14, where *Central Mercantile Residences (PG3)* exhibit a positive loading on both RC2 (accessibility to amenities) and RC4 (walkability and inter-neighbourhood connectivity), whereas *Countryside Marketplace Vistas (PG4)* show otherwise.

## Conclusion

To objectively evaluate the influence of precinct design on residents' perceived QoL, this chapter has introduced a multi-dimensional technique to classify precinct typologies, based on precinct attributes defined by planning parameters. These precinct typologies could then be used to examine, through statistical analysis, whether residents residing in different precinct types exhibit different perceived QoL. This multi-dimensional technique was demonstrated and discussed in the context of three public housing residential towns in Singapore, illustrating its real-world application for scholars in the spatial planning and design fields.

The case study in Punggol has adeptly exemplified the effective application of the proposed multi-dimensional technique, coupled with statistical analysis, for discerning different levels of residents' perceived QoL within specific precinct types. These varying QoL perceptions extend to various facets, including sense of belonging, satisfaction with the physical and built environment, receipt of support from neighbours, and contentment with mobility and transport. These findings substantiate the relationship between good-quality physical environments, particularly spatial attributes associated with social dimensions, and the positive impacts experienced by residents (Dempsey, 2008). The outcomes obtained from such quantitative methodology are important indicators to guide further qualitative research, using 'thick data' to understand underlying trends, uncovering the human stories, emotions, and motivations at play (Wang, 2016). When qualitative research is combined with a quantitative multi-dimensional technique in an integral manner, a more comprehensive understanding of human dynamics in the urban systems can be obtained (Chong et al., 2022).

While the MDC has effectively identified different precinct types, respectively, in Toa Payoh and Jurong East, the statistical analysis revealed a lack of statistically significant differences in the QoL among residents residing in different precinct types that consist of precincts primarily constructed during the 1970s and 1980s. This necessitates further study to examine whether a homogenising effect exists in both Toa Payoh and Jurong East, wherein residents of longer residency may find ways to compensate what were originally perceived as shortcomings in the physical planning. This also suggests the need for investigating other latent or relational variables, e.g., social fabrics, in understanding and comparing residents' perceived QoL, especially for residents with longer residency.

In conclusion, this chapter offers a guiding technique for a data-driven, multi-dimensional approach to understanding the complex interplay between the multi-faceted precinct design and residents' perceived QoL in high-rise, high-density urban living, such as the public housing neighbourhoods in Singapore. It also underscores the importance of factoring in the social fabric within public housing residential towns. One limitation of the technique is that precinct comparison is currently only made within the individual town, and not across town, as the principal components that are used to classify the precinct typologies are unique to each town. More studies are thus needed to identify planning parameters that are nuanced yet sufficiently general for all towns, in order to overcome this limitation. Future research could also delve into the impacts of precinct design, including the influence of covered linkway layouts and block morphologies on different levels of neighbourliness, as suggested in the Punggol case. This chapter aimed to serve as an impetus for a more comprehensive and nuanced understanding of these intricate people-environment relationships, with the potential to inform future urban planning decisions and design, especially in high-density urban living environments.

## Bibliography

Alexiou, A., Singleton, A., & Longley, P. A. (2016). A classification of multidimensional open data for urban morphology. *Built Environment, 42*(3), 382 395. https://doi.org/10.2148/benv.42.3.382

Bouguettaya, A., Yu, Q., Liu, X., Zhou, X., & Song, A. (2015). Efficient agglomerative hierarchical clustering. *Expert Systems with Applications, 42*(5), 2785–2797. https://doi.org/10.1016/j.eswa.2014.09.054

Brown, R. B., Dorius, S. F., & Krannich, R. (2005). The boom-bust-recovery cycle: Dynamics of change in community satisfaction and social integration in Delta, Utah*. *Rural Sociology, 70*, 28–49.

Chan, E. H. W., & Lee, G. K. L. (2009). Design considerations for environmental sustainability in high density development: A case study of Hong Kong. *Environment, Development and Sustainability, 11*(2), 359–374. https://doi.org/10.1007/s10668-007-9117-0

Cho, M., Ha, T. M., Lim, Z. M. T., & Chong, K. H. (2018). "Small places" of ageing in a high-rise housing neighbourhood. *Journal of Aging Studies, 47*, 57–65. https://doi.org/10.1016/j.jaging.2018.10.003

Chong, K. H., Ha, T. M., To, K., & Chau, Y. (2022). Inclusive smart community-towards a socially integrated and enabling community. In *Future urban habitation: Transdisciplinary perspectives, conceptions, and designs* (pp. 49–66). Chichester: Wiley Blackwell. https://doi.org/10.1002/9781119734895.ch3

Chong, K. H., & Tan, D. (2015). *Designing communities: Evolutions of precinct public space design in Singapore public housing estates*. Gwanju: East Asian Architectural Culture.

Cooper, E., Gates, S., Grollman, C., Mayer, M., Davis, B., Bankiewicz, U., & Khambhaita, P. (2019). *Transport, health, and wellbeing: An evidence review for the Department for Transport*. London: NatCen Social Research.

Daniels, R., & Mulley, C. (2013). Explaining walking distance to public transport: The dominance of public transport supply. *Journal of Transport and Land Use, 6*(2), 5–20. https://doi.org/10.5198/jtlu.v6i2.308

Day, W. H. E., & Edelsbrunner, H. (1984). Efficient algorithms for agglomerative hierarchical clustering methods. *Journal of Classification, 1*(1), 7–24. https://doi.org/10.1007/BF01890115

Dempsey, N. (2008). Does quality of the built environment affect social cohesion? *Proceedings of the Institution of Civil Engineers-Urban Design and Planning, 161*(3), 105–114. https://doi.org/10.1680/udap.2008.161.3.105

Dempsey, N. (2009). Are good-quality environments socially cohesive? Measuring quality and cohesion in urban neighbourhoods. *The Town Planning Review, 80*(3), 315–345.

Department of Statistics Singapore. (2020). *Population trends, 2020*. https://www.singstat.gov.sg/-/media/files/publications/population/population2020.pdf

Filkins, R., Allen, J. C., & Cordes, S. (2000). Predicting community satisfaction among rural residents: An integrative model. *Rural Sociology, 65*(1), 72–86.

Fleury-Bahi, G., Félonneau, M.-L., & Marchand, D. (2008). Processes of place identification and residential satisfaction. *Environment and Behavior, 40*(5), 669–682. https://doi.org/10.1177/0013916507307461

Gil-Begue, S., Bielza, C., & Larrañaga, P. (2021). Multi-dimensional Bayesian network classifiers: A survey. *Artificial Intelligence Review, 54*(1), 519–559. https://doi.org/10.1007/s10462-020-09858-x

Grannis, R. (2009). *From the ground up: Translating geography into community through neighbour networks* (pp. 17–27). Princeton: Princeton University Press.

Grzeskowiak, S., Sirgy, M., & Widgery, R. (2003). Residents' satisfaction with community services: Predictors and outcomes. *The Journal of Regional Analysis and Policy, 33*, 1–36.

Heng, C. K. (2017). *50 Years of urban planning in Singapore*. Singapore: World Scientific.

Hidalgo, M. C., & Hernandez, B. (2001). Place attachment: Conceptual and empirical questions. *Journal of Environmental Psychology, 21*(3), 273–281. https://doi.org/10.1006/jevp.2001.0221

Housing & Development Board (HDB). (1995). *Social aspects of public housing in Singapore: Kinship ties and neighbourly relations*. Singapore: Research Section, Research & Planning Dept., Housing & Development Board.

Housing & Development Board (HDB). (2018). *HDB introduces town design guides to strengthen the distinctive identities of 24 HDB towns*. Retrieved from https://www.nas.gov.sg/archivesonline/data/pdfdoc/20180904001/Press%20release%20-%20HDB%20Town%20Design%20Guide%20-%2004092018.pdf

Housing & Development Board (HDB). (2020a). *Jurong East*. Retrieved from https://www.hdb.gov.sg/about-us/history/hdb-towns-your-home/jurong-east

Housing & Development Board (HDB). (2020b). *Punggol*. Retrieved from https://www.hdb.gov.sg/about-us/history/hdb-towns-your-home/punggol

Ibes, D. C. (2015). A multi-dimensional classification and equity analysis of an urban park system: A novel methodology and case study application. *Landscape and Urban Planning, 137*, 122–137. https://doi.org/10.1016/j.landurbplan.2014.12.014

Kamalipour, H., Yeganeh, A. J., & Alalhesabi, M. (2012). Predictors of place attachment in urban residential environments: A residential complex case study. *Procedia – Social and Behavioral Science, 35*, 459–467. https://doi.org/10.1016/j.sbspro.2012.02.111

Kim, J. (2017). Comparing the influences of the D/H ratio, size, and facade design of an enclosed square on its perceptual qualities as a sustainable urban space in South Korea. *Sustainability, 9*(4). https://doi.org/10.3390/su9040675

Kruskal, W. H., & Wallis, W. A. (1952). Use of ranks in one-criterion variance analysis. *Journal of the American Statistical Association, 47*, 583–621. https://doi.org/10.2307/2280779

Lane, A. P., Hou, Y., Hooi Wong, C., & Yuen, B. (2020). Cross-sectional associations of neighborhood third places with social health among community-dwelling older adults. *Social Science & Medicine, 258*. https://doi.org/10.1016/j.socscimed.2020.113057

Menz, S. (2014). Contexts—Increasing density: Elevated spaces in Singapore. In S. Menz (Ed.), Public space evolution in high-density living in Singapore: Ground and elevated public spaces in public housing precincts (pp. 25–28). Singapore: Singapore-ETH Centre (SEC) Future Cities Laboratory (FCL).

Murtagh, F., & Contreras, P. (2012). Algorithms for hierarchical clustering: An overview. *WIREs Data Mining and Knowledge Discovery, 2*(1), 86–97. https://doi.org/10.1002/widm.53

OneMap. (2018). OneMap: *The authoritative national map of Singapore with the most detailed and timely updated information.* Singapore Land Authority, SLA. Retrieved from https://www.onemap.gov.sg/main/v2/

Ooi, G. L., & Tan, T. T. W. (1992). The social significance of public spaces in public housing estates. In B. H. Chua & N. Edwards (Eds.), *Public space: Design, use and management* (pp. 69–81). Singapore: Singapore University Press.

Read, J., Bielza, C., & Larrañaga, P. (2014). Multi-dimensional classification with super-classes. *IEEE Transactions on Knowledge and Data Engineering, 26*(7), 1720–1733. https://doi.org/10.1109/TKDE.2013.167

Singapore Land Authority (SLA). (2021, January 25). *Land area and dwelling units by town.* Retrieved from Data.Gov.Sg. https://data.gov.sg/dataset/land-area-and-dwelling-units-by-town?view_id=ef3d1ad0-61fe-413d-a165-92b9670fc00f&resource_id=898d9 85a-0996-4efd-b2c2-7d9fab4138e9

Urban Redevelopment Authority (URA). (1995). *Toa Payoh planning area: Planning report 1995.* Urban Redevelopment Authority (URA).

Wang, T. (2016, January 20). *Why big data needs thick data.* Medium. Retrieved from https://medium.com/ethnography-matters/why-big-data-needs-thick-data-b4b3e75e3d7

# 4

# URBAN MOBILITY AND NEIGHBOURHOOD ACTIVITY ANALYSES USING DATA FROM SMART MOBILE DEVICES

*Zann Koh, Hasala Marakkalage, Billy Pik Lik Lau, Yuren Zhou, Ran Liu and Chau Yuen*

## Introduction

To facilitate Smart Community Design, knowledge about the current state of urban human mobility within the relevant community, i.e., people within the community and the surrounding environment, must first be gathered and analysed before further planning decisions may be made. If there is a selected urban area for upgrading, knowing how the local community of people living there currently move around and utilise the space is of paramount importance to facilitate informed decision-making as well as to ensure that any planned human flow around the space is safe and efficient.

Some challenges faced in this process include gathering of accurate data from multiple sources as well as the lack of a structured framework to analyse the gathered data to extract insights to aid in the making of informed decisions when planning a community. Therefore, this chapter presents selected methods of data analysis performed on a combination of Wi-Fi and GPS data. This chapter will be split into two parts as follows.

The first part of the chapter introduces a method for analysis of points of interest (POIs) for individual users. This proposed method makes use of both Wi-Fi and GPS data from each user's mobile devices to identify the distinct Wi-Fi-based POI of users in an indoor environment, the revisited POI by the same set of users, and the common POI among users. Following that, the neighbourhood activity is identified, and a heat map can be plotted by excluding stay home duration. Subsequently, the neighbourhood micro mobility patterns that move under covered walkways or cut across buildings can also be identified using data fusion.

The second part presents a new mobility metric for use with GPS data linked to individuals. The data for this study was collected from a group of volunteers via

DOI: 10.4324/9781003437659-7

a user-installed smartphone application. This metric is shown to be different for Working and Non-working users on Workdays and Offdays, respectively. The new mobility metric is used together with a clustering algorithm to profile a group of users based on their travel patterns and tendency to travel to each distance range. Afterwards, the cluster results are analysed using two other proposed analysis metrics to identify common characteristics of users in the same cluster as well as common types of POI visited by users in each cluster.

## Urban Mobility Analysis via Clustering of Wi-Fi Signals

### Case Study 1: Background

To develop a better understanding of users' current needs and patterns for urban planning, it is crucial to first gather relevant data. Mobile crowdsensing (MCS) has garnered a large amount of attention due to three aspects of smart mobile devices – their pervasiveness in a large proportion of the population, their sensing capabilities, as well as them being carried around everywhere by humans. Examples of the myriad MCS applications in various sectors include transportation (Farkas, Feher, Benczur, & Sidlo, 2015), healthcare (Leonardi, Cappellotto, Caraviello, Lepri, & F. Antonelli, 2014), and social networking platforms (Hu, Li, Ngai, Leung, & Kruchten, 2014). There have been phenomena that were monitored by diverse information harnessed through smartphone applications with proper crowd participation (Hoteit, Secci, Sobolevsky, Ratti, & Pujolle, 2014; Kang, Sobolevsky, Liu, & Ratti, 2013). In mobility analysis applications, identifying detailed motion pattern information (outdoor and indoor) provides comprehensive insights on user mobility (Li, Zheng, Xie, Chen, Liu, & Ma, 2008; Lou, Zhang, Zheng, Xie, Wang, & Huang, 2009; Gamanayake, Jayasinghe, Ng, & Yuen, 2020; Helgason, Kouyoumdjieva, & Karlsson, 2013). Knowing the user POIs is paramount in mobility tracking applications to provide relevant urban and facility planning for local users. Data obtained from smartphones have been used in several applications, such as detecting indoor or outdoor environments with data fusion (Zhou, Zheng, Li, Li, & Shen, 2012; Lau, Hasala, Kadaba, Thirunavukarasu, Yuen, Yuen, & Nayak, 2017; Shin, Chon, & Cha, 2012) as well as the understanding of elderly lifestyles (Marakkalage, Sarica, Lau, Viswanath, Balasubramaniam, Yuen, Yuen, Luo, & Nayak, 2018). Nonetheless, these works are insufficient to identify the indoor POI granularity in contemporary urban places such as shopping malls and high-rise residential complexes.

Mobility tracking in indoor environments, especially in high-rise urban buildings and apartments, is challenging. Using only GPS data is insufficient for identifying when people leave their homes and visit common areas within the same building or neighbourhood POI. This is due to the two-dimensional nature of GPS data, which limits differentiation between multiple indoor POIs. In this section, the scope of indoor POI identification focuses on the mobility pattern of a typical user in commonly visited indoor environments like shopping malls or apartment complexes with multiple POIs in a single building. The GPS data of a single user may get clustered into one POI when visiting one shopping mall when the user may have visited multiple POIs such as visiting different shops within the same mall.

Incorporating Wi-Fi data can help to distinguish between different locations in the same building even when the GPS data accuracy is low, since urban environments in recent times contain plenty of Wi-Fi access points (APs). Hence, by combining or fusing GPS and Wi-Fi information, it is possible to identify indoor POIs (as first introduced in a previous work (Marakkalage, Liu, Viswanath, & Yuen, 2019) and improved POI extraction technique later in this section). Prior research has utilised Wi-Fi AP information to generate indoor floorplans (Shin, Chon, & Cha, 2012; Alzantot & Youssef, 2012) and to identify indoor locations through localisation (Zhu, Zheng, Xu, & Li, 2014; Liu, Marakkalage, Padmal, Shaganan, Yuen, Guan, & Tan, 2019a, 2019b; Liu, Yuen, Do, & Tan, 2017; Tian, Wu, Li, & Wang, 2019). A major drawback of those works is that high power consumption is incurred due to high sampling rate requirements, which is a prime challenge in MCS (Ganti, Ye, & Lei, 2011; Lau, Marakkalage, Zhou, Hassan, Yuen, Zhang, & Tan, 2019; Wu, Yang, & Liu, 2014; Marakkalage, Lau, Viswanath, Yuen, & Yuen, 2019). Another drawback is that the creation of indoor fingerprint maps involves high labour cost. The data collection and preprocessing of Wi-Fi and GPS data for this study is presented in the subsection "System Overview", while the POI extraction process and analysis is covered in "Indoor POI Study".

Obtaining the identified POI paves the way for further analysis of neighbourhood activities and micro mobility within a neighbourhood. Neighbourhood activity analysis is conducted to understand the POIs where users visit in their neighbourhood of residence. These POIs include common areas within high-rise residential estates, such as convenience stores on the ground floor, void decks, or sky gardens with benches for intermingling with friends and neighbours. The fusion of Wi-Fi data is essential here to adapt to the lack of vertical mobility coverage from GPS data. Details and analysis are covered in the subsection "Neighbourhood Activity Study".

Lastly, the subsection "Micro Mobility Study" presents the analysis of common routes taken by multiple users between POIs. The Wi-Fi fingerprints are clustered to identify the movement patterns that may be obscured from GPS data under covered walkways or cutting across buildings.

## System Overview

Analysing the mobility of a user starts with identifying the trajectories that they take. A sample trajectory of a user is shown in Figure 4.1. The raw data is first processed to extract GPS stay points. Next, Wi-Fi data is incorporated to extract POIs that represent the indoor POI visited by the user, the activity performed around neighbourhood POIs by the user, and finally the micro mobility, which are the links

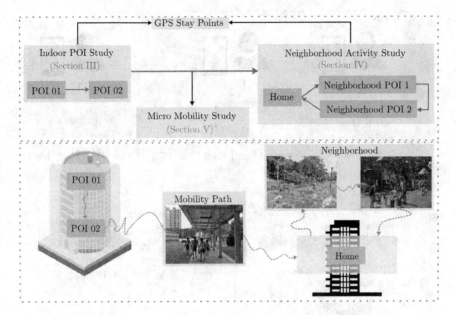

**FIGURE 4.1**    Example of a user's trajectory.

*Source*: Authors.

between the POIs that are studied. These three aspects of a user's trajectory are explored in this study.

## Data Collection

Data for this study was collected via a smartphone application (front-end) to GPS location data, and surrounding Wi-Fi AP information. The data was then transferred to a cloud-based server application in the back end for further processing. An overview of the data flow of the system is shown in Figure 4.2.

According to the Android API (Google, 2018), excessive scanning of Wi-Fi and GPS has a large impact on mobile phone battery consumption. Hence, the smartphone application scans the surrounding Wi-Fi AP information, the Media Access Control (MAC) address, and corresponding Received Signal Strength (RSS) only once every five minutes.

The Wi-Fi scanning process is described as follows. The MAC address of the Wi-Fi AP is denoted as m, while the RSS of the AP is denoted as $r$ in units of dBm. The scan result(s), which is the list of surrounding AP MAC addresses and their corresponding RSS, are shown in Equation 1, where $n$ is the number of APs observed in each scan result.

$$s = \left\{ m_1, r_1 \right\}, \left\{ m_2, r_2 \right\}, \ldots, \left\{ m_n, r_n \right\}$$

(1)

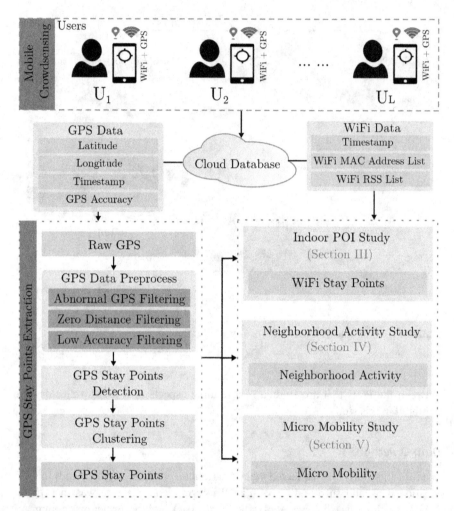

**FIGURE 4.2**    Overview of the proposed system.

*Source*: Authors.

Each scan result and the corresponding timestamp ($t$) of the Wi-Fi scan is stored in a list of scan results ($S$), denoted as shown in Equation 2, where $m$ is the number of scan results in $S$.

$$S = \{s_1, t_1\}, \{s_2, t_2\}, \ldots, \{s_m, t_m\} \tag{2}$$

This list of scan results is then stored locally in the device until it is uploaded to the back end for further analysis. The raw data is compressed before uploading to reduce the cost of data transmission to the back end. A duration of six hours is selected for the upload interval as it has a significant reduction in size when compressed. The data is only uploaded when the user's device is connected to a Wi-Fi network. Otherwise, the data is stored locally while waiting to connect to a network. A total of 12 users participated in the study and will be referred to with the alphabets from A to L.

## Data Processing

After the data points are received in the back end, the GPS stay points are first extracted using the stay point extraction module within the data processing pipeline, as shown in Figure 4.2. First, the raw GPS data undergoes processing such as the removal of abnormal data, sequences with zero distance, as well as low accuracy GPS locations (Marakkalage, Lau, Viswanath, Yuen, & Yuen, 2019). The abnormal data here includes GPS data with sudden location shifts within a short period of time, which can distort the actual path travelled by the users. Zero distance often occurs when the GPS does not receive any signal from the satellite, which causes the exact same location for subsequent data. This does not provide any meaningful data and hence it is filtered out. The next data processing technique is accuracy filtering, where low accuracy GPS data that causes high uncertainty in determining the actual location of users is removed. Subsequently, GPS stay point extraction (Lau, Hasala, Kadaba, Thirunavukarasu, Yuen, Yuen, & Nayak, 2017; Zheng, Zhang, Xie, & Ma, 2009) is performed to obtain the list of POIs from a particular user with the timestamp for each visit. Afterwards, the POIs are clustered based on their geographical location using DBSCAN. This briefly explains the GPS stay points extraction method. After obtaining the raw GPS stay points, the durations of the GPS stay points are used to further detect indoor POI for the users.

## Indoor POI Study

For the indoor POI study, the techniques that are used to extract indoor POI of the users, by processing the GPS and Wi-Fi data collected through the smartphone-based mobile application, are presented. Indoor POIs are extracted by clustering the Wi-Fi fingerprints and matching the corresponding cosine similarity scores. Table 4.1 shows the symbols used in this section, and their description for the convenience of the reader.

To choose a suitable clustering algorithm, different clustering algorithms that have been experimentally evaluated by previous research (Lau, Hasala, Kadaba, Thirunavukarasu, Yuen, Yuen, & Nayak, 2017) were examined, and DBSCAN (Ester, Kriegel, Sander, Xu, et al., 1996) was chosen as the most suitable method because of its ability to form arbitrary shaped clusters. A modified DBSCAN algorithm to cluster the Wi-Fi RSS measurements is thus used to cluster the Wi-Fi RSS measurements. Cosine similarity scores between two RSS values are used as the distance metric of the modified DBSCAN algorithm. This study defines a cluster (POI) to be formed when a user stays for at least 20 minutes in the same place.

**TABLE 4.1** Symbols and their description for clustering algorithm

| Symbol | Description |
|--------|-------------|
| $\epsilon$ | Similarity threshold |
| $F$ | Wi-Fi fingerprint |
| $R$ | RSS (average) in dBm |
| $P$ | Distinct MAC address count |
| $D$ | Cosine similarity distance |
| $\alpha, \beta$ | Scan result from a list of scan results |
| $Y$ | Dot product of two Wi-Fi fingerprints |
| $\Omega$ | Number of common mac addresses |
| $C$ | Cosine similarity score |

*Source*: Authors.

**TABLE 4.2** Impact of cosine similarity threshold for indoor POI Extraction

| Ground Truth | Start Time (hh:mm) | End Time (hh:mm) | POI ID | |
|--------------|--------------------|--------------------|------------------------|------------------|
| | | | $\epsilon = adaptive$ | $\epsilon = 0.5$ |
| Home | 00:00 | 09:23 | 01 | 01 |
| Office | 09:59 | 11:34 | 02 | 02 |
| Meeting room | 11:58 | 14:57 | 03 | 03 |
| Canteen | 15:29 | 16:39 | 04 | 04 |
| Office | 16:44 | 17:09 | 02 | 02 |
| Home | 17:49 | 18:54 | 01 | 01 |
| Home | 18:59 | 20:39 | 01 | 05 |
| Home | 20:48 | 23:53 | 01 | 01 |

*Source*: Author.

For DBSCAN, some parameters are required to be selected beforehand such as the minimum required points to form a cluster (minPts) and the cosine similarity threshold ($\epsilon$). The parameter selection of $\epsilon$ between an adaptive and a fixed value is done by experimental evaluation as presented in Table 4.2. The performance of both threshold values is evaluated together with the ground truth labels for the user C. When the threshold is adaptive, the indoor POI extraction result aligns with the ground truth. An additional POI ID 05 is identified when the threshold value is 0.5 as highlighted in red in Table 4.2. This is due to a smaller Wi-Fi AP list size ($A_L$), which is less than 35, observed in residential environments as compared to the shopping mall or office environments. The result in Table 4.2 shows that when the similarity threshold is fixed, even the changes in the size of the scanned AP list have an impact on the Wi-Fi cluster formation.

The final selected parameter settings are minPts = 4 and $\epsilon$ is set as adaptive.

Algorithm 1 explains the procedure of clustering for a given set of Wi-Fi data ($S$), the similarity threshold $\epsilon$, and the minimum points to form a cluster minPts.

$P$ is the list of output cluster points. The process of obtaining the neighbour points is shown in Algorithm 2, where inputs are $\alpha$ and $S$, and the output is $N$, which are mentioned in Algorithm 1. Algorithm 3 covers the similarity threshold used for determining neighbour points in Algorithm 2.

---

**Algorithm 1** POI extraction from raw WiFi data

---

**Input:** similarity threshold $(\epsilon)$, $minPts$, WiFi list $(S)$
**Output:** Cluster point list $(P)$
Visited points $(V_p)$, index $(z_1)$, $P = 0$
while *size of $S \geq z_1$* do
   $\alpha = S[z_i]$
    if $\alpha \not\subset V_p$ then
     add $\alpha$ to $V_p$
     $N$ = get neighbours of $\alpha$
     if *size of $N \geq minPts$* then
      $z_2 = 0$
      while *size of $N \geq z_2$* do
       $\beta = N[z_2]$
       if $\beta \not\subset V_p$ then
        add $\beta$ to $V_p$
        $Q$ = get neighbours of $\beta$
        if *size of $Q \geq minPts$* then
         merge $Q$ with $N$

        end if
       else
        $z_2 = z_2 + 1$

       end if
      end while
      add $N$ to $P$
     end if
   else
    $z_1 = z_1 + 1$

   end if
end while

---

---

**Algorithm 2** Obtaining the Neighbour Points

---

**Input:** Scan result $(\alpha)$, WiFi list $(S)$
**Output:** Neighbour points $(N)$
$N = 0$
for *every index i in S* do
   $D$ = calculate similarity$(\alpha, S[i])$
   $\epsilon$ = calculate threshold$(\alpha, S[i])$
   if $D \geq \epsilon$ then
    add $S[i]$ to $N$
   end if
end for

---

---
**Algorithm 3** The process of threshold calculation

---
**Input:** Fingerprints $(F_1, F_2)$
**Output:** Similarity threshold $(\epsilon)$
**if** $F_1 \leq A_L$ and $F_2 \leq A_L$ **then**
 |  $\epsilon = \epsilon_L$
**else**
 |  $\epsilon = \epsilon_H$

**end if**

---

After getting the final list of clusters, a fingerprint for each cluster (indoor POI) is generated with a unique POI ID (i.e., indoor POI ID is unique to a given GPS stay point). The POI fingerprint ($F$) is denoted as shown in Equation 3, where $M$ is the MAC address of a cluster, $R$ is the corresponding average RSS in dBm, and $p$ is the number of distinct MAC addresses scanned at that POI.

$$F = \left\{ M_1, R_1 \right\}, \left\{ M_2, R_2 \right\}, \ldots, \left\{ M_p, R_p \right\} \tag{3}$$

Cosine similarity is used as the distance metric in the DBSCAN algorithm for this case study. As an example, two Wi-Fi fingerprints $F_1$ and $F_2$ are shown below:

$$F_1 = \left\{ M_1^1, R_1^1 \right\}, \left\{ M_2^1, R_2^1 \right\}, \ldots, \left\{ M_u^1, R_u^1 \right\} \tag{4}$$

$$F_2 = \left\{ M_1^2, R_1^2 \right\}, \left\{ M_2^2, R_2^2 \right\}, \ldots, \left\{ M_v^2, R_v^2 \right\} \tag{5}$$

where $u$ and $v$ denote the number of distinct MAC addresses in $F_1$ and $F_2$, respectively. To obtain the cosine similarity, the first step is to calculate the dot product of RSS in common MAC addresses for the two fingerprints ($Y$). This is shown in Equation 6, where $w$ denotes the number of common MAC addresses.

$$Y = \sum_{i=1}^{w} \left[ R_i^1 \cdot R_i^2 \right] \tag{6}$$

Next, the dot products of each RSS in $F_1$ and $F_2$ are calculated according to Equations 7 and 8, respectively.

$$d_1 = \sum_{j=1}^{u} \left[ R_j^1 \cdot R_j^1 \right] \tag{7}$$

$$d_2 = \sum_{k=1}^{v} \left[ R_k^1 \cdot R_k^1 \right] \tag{8}$$

Finally, the cosine similarity ($C$) between the two Wi-Fi fingerprints is calculated according to Equation 9.

$$C = Y / \left( \sqrt{d_1} \times \sqrt{d_2} \right); \text{ where } 0 \leq C \leq 1 \tag{9}$$

To verify the accuracy of the single-user POI identification, the Wi-Fi clustering results are compared along with the ground truth. Table 4.3 presents the single-user POI identification for user H, over a duration of one year of POI visits in Changi General Hospital. According to Table 4.3, the POI ID is different for different locations (i.e., Wi-Fi clusters) inside the building. This shows that the single-user POI identification technique can detect when the user revisits POIs with the same ID 04, 07, 10, 11, and 12, on different dates of the year.

The next step after verification of POI for a single user is the analysis of the popular POI among users. The Louvain method for community detection (Blondel, Guillaume, Lambiotte, & Lefebvre, 2008) is utilised to gain insights on popular POI among users. In each indoor environment, let the number of POI be $h$, and the number of pair-wise cosine similarity scores ($I$) is calculated according to Equation 10.

$$I = \frac{h!}{2!(h-2)!} \tag{10}$$

The Louvain algorithm takes $I$ as the input, obtains the optimum partitioning among POI (nodes) by comparing pair-wise similarity (edges) scores, and provides the modularity score as the output. The results of indoor POI extraction are presented as follows.

The location of study of popular indoor POIs among multiple users is selected to be Changi City Point (CCP), where 11 users visit for the purpose of shopping or dining over a duration of three months. The clustering results detected 41 indoor

**TABLE 4.3** Single-user indoor POI results

| Ground Truth | Date (yyyy-mm-dd) | Start Time (hh:mm) | End Time (hh:rnm) | POI ID |
|---|---|---|---|---|
| Centre for | 2019-01-22 | 11:32 | 11:57 | 04 |
| Innovation | 2019-02-21 | 16:09 | 17:36 | |
| | 2019-08-06 | 10:02 | 10:52 | |
| Level 6 room | 2019-04-02 | 10:58 | 12:23 | 06 |
| Main board room | 2019-05-08 | 16:58 | 18:48 | 07 |
| | 2019-08-13 | 15:10 | 15:35 | |
| Level 8 room | 2019-07-11 | 09:44 | 10:24 | 10 |
| | 2019-07-18 | 08:39 | 09:32 | |
| Level 7 room | 2019-07-24 | 15:14 | 16:14 | 11 |
| | 2019-07-31 | 08:48 | 09:59 | |
| | 2019-08-01 | 11:57 | 14:14 | |
| Ward 45 | 2019-08-02 | 13:12 | 15:54 | 12 |
| | 2019-10-15 | 15:15 | 15:54 | |

*Source*: Authors

POIs at the CCP. Substituting into Equation 10, a value of $I = 820$ pair-wise similarities are calculated for the input into the Louvain algorithm. Table 4.4 shows the modularity scores calculated for different similarity thresholds for partitioning to understand the optimum partition for communities. A higher modularity score indicates a better partitioning of communities.

Different POIs may differ in terms of area of coverage, for example, a food court is larger than a clothing shop. The objective of this study is to detect even the smallest POI visited by users. Therefore, for POI identification, a threshold value of 0.5, with the highest similarity and a higher modularity score than the threshold value of 0.2, was selected as the partitioning threshold for community detection.

Table 4.5 shows the details of the common POI visited by the 11 users in CCP. The shopping mall is a three-storey building with Basement 1 (B1), Level 1 (L1), and Level 2 (L2). Users from the study carried on with their normal routine to the mall for shopping or dining purposes.

**TABLE 4.4** Louvain modularity score for different partitioning thresholds

| Threshold Value | Modularity Score |
| --- | --- |
| 0.2 | 0.625 |
| 0.3 | 0.803 |
| 0.4 | 0.766 |
| 0.5 | 0.692 |

*Source*: Authors.

**TABLE 4.5** Common POI among different users in CCP Shopping Mall

| Floor Level | Ground Truth | POI ID | User(s) |
| --- | --- | --- | --- |
| B1 | Restaurant 1 | 00 | H |
| | Restaurant 2 | 11 | J |
| | Restaurant 3 | 29 | K |
| | Drink shop | 05 | G |
| | Utility store | 13 | K |
| L1 | Restaurant 4 | 11 | B, H |
| | Clothing shop 1 | 27 | A |
| L2 | Clothing shop 2 | 08 | C |
| | Clothing shop 3 | 09 | C, D, I |
| | Food court | 01, 02 | A |
| | | 03, 05 | C |
| | | 07, 10 | D |
| | | 12, 15 | F |
| | | 21, 23 | J |
| | | 24, 25 | K |

*Source*: Authors.

For common POIs, it can be observed that Restaurant 4 was visited by two users while the other restaurants were visited by a single user each. Clothing Shop 3 was visited by more users than the other clothing shops. Since the sample size for this study is small, it is not representative of the shopper population in general. However, this also means that obtaining more representative data and insights would be possible with a larger sample size.

One possible limitation observed was that there were three different restaurants in B1 that were visited by users, with the indoor POI IDs 00, 11, and 29, respectively. Restaurant 4 in L1 was assigned the same indoor POI ID (i.e., POI ID 11) as Restaurant 2 in B1, where user J visited. Upon checking the layout of the shopping mall, the observation was made that Restaurant 4 was located directly above Restaurant 2, as shown in Figure 4.3. This, combined with the large distance between them, could have caused their Wi-Fi measurements to be similar.

Another observation that can be made is that the food court on L2 is divided into multiple POIs. This could be due to the food court covering a large area, almost half of L2, and the Wi-Fi RSS measurements fluctuating due to large crowds. These observations indicate that for further studies, an emphasis on having accurate ground truth is crucial for comparison of results.

With more specific insights drawn from these indoor POIs, this can facilitate Smart Community Design by showing the types of common POIs visited by multiple users. An analysis into the relative location of each indoor POI may also show some relation between distance between POIs with similar purposes (different clothes shops, different restaurants) and frequency of visits by the same user on the same trip.

**FIGURE 4.3** CCP Basement 1 and Level 1 layout.

## Neighbourhood Activity Study

In an urban area, most of the residents tend to visit nearby places of home or office for shopping and leisure activities during their free time. Especially in dense areas, where high-rise buildings are common as residences, conventional GPS clustering approach using GPS data may indicate such a building as one POI, but, in reality, there are many possible POIs (e.g., convenient store, common area, BBQ pit, etc.) in a multistorey setting. This is due to the dimensionality nature of GPS data, and GPS data alone cannot provide accurate information on stay points at a micro level. Moreover, it is useful to understand the user's stay points in the residential (home) neighbourhood. We define such stay points or places of short duration as neighbourhood activity, and exploit Wi-Fi fingerprints along with GPS data to identify such neighbourhood activity.

To extract the neighbourhood activity from the trajectory, the concept of sensor fusion is leveraged to combine GPS and Wi-Fi information sources. The overall process of the neighbourhood activity extraction is illustrated in Figure 4.4.

The two main data sources used in the processing stage are Wi-Fi and GPS stay points. First, the GPS stay points are identified and labelled according to the

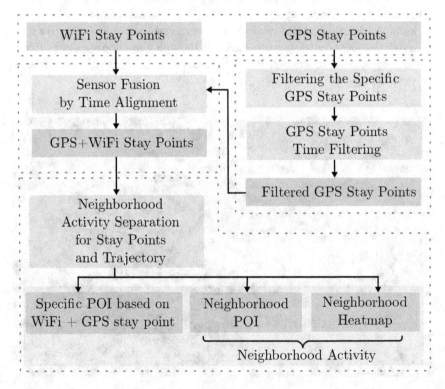

**FIGURE 4.4**   Neighbourhood activity data processing pipeline.

characteristics of each GPS POI. Subsequently, the POIs are filtered by time to focus on specific POIs. Examples of specific POIs include a user's home location or office location. Next, Wi-Fi stay points are generated according to Algorithm 1. The filtered GPS stay points and the Wi-Fi stay points are then fused to generate GPS + Wi-Fi stay points. Home and office stay points can be deduced heuristically by stay duration as both are stay points with the longest stay durations. The rest of the GPS + Wi-Fi stay points are categorised as neighbourhood POI. The remaining raw GPS points (moving points) that occurred between neighbourhood POI and a specific POI can be further converted into heat maps to capture potential neighbourhood activity that does not form a stay point.

As proof of concept, a simple case study for user H was performed using Wi-Fi and GPS data for a 6-hour period of the day of 18 September 2019. A comparison of the detected POIs from GPS and GPS + Wi-Fi stay points is illustrated in Figure 4.5a. It is shown that the neighbourhood stay points are accurately detected using the GPS + Wi-Fi stay points data fusion method, whereas the GPS-only method is insufficient.

Next, a visualisation of the stay points is shown in Figure 4.5b. The grey icons represent travelling GPS data, while the pink and green icons denote home and neighbourhood POI, respectively. The neighbourhood area is located less than 100 m away from the residential area. Using only GPS stay points, it may appear that there is a high probability of the neighbourhood activity area being the same stay point as the residential area due to their proximity. Hence, the GPS + Wi-Fi method is more accurate in distinguishing nearby stay points.

After the benefit of using GPS + Wi-Fi stay point detection method was confirmed, a more in-depth study was performed on the data of user H over eight

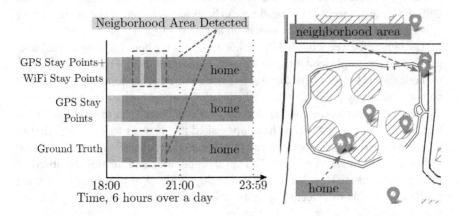

(a) Stay points extraction methods comparison

(b) Neighborhood activity's visualization

FIGURE 4.5   Toy example of performing neighbourhood activity extraction.

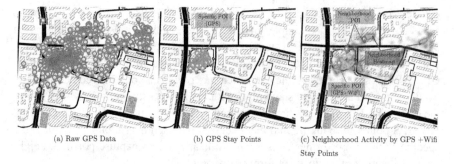

(a) Raw GPS Data          (b) GPS Stay Points          (c) Neighborhood Activity by GPS +Wifi Stay Points

**FIGURE 4.6**    Comparison of raw data, GPS stay points, and neighbourhood activity by GPS + Wi-Fi stay points, for user H, from 01 April 2019 to 01 December 2019.

*Source*: Author.

months starting from 01 April 2019 to 01 December 2019. The raw GPS data surrounding the home location of user H is extracted. The same temporal notion of the home stay point is applied to Wi-Fi data to study neighbourhood activity.

Figure 4.6 shows the comparison of neighbourhood activity obtained by GPS and GPS + Wi-Fi methods. Figure 4.6a plots the raw data that user H has travelled to nearby places from home. The GPS-only stay point detection method applied in Figure 4.6b fails to detect that the user has travelled to nearby places. This could have happened because the POI user travelled to a nearby location, which is indistinguishable by using GPS data only. Using the GPS + Wi-Fi stay points in Figure 4.6c, it is possible to detect other neighbourhood POIs that the user has visited (green, blue, and purple icons). Additionally, since Wi-Fi data is fused with GPS stay points, specific POI locations can be exactly identified, and the remaining moving raw GPS points can be converted to heat maps.

The blue and green icons in Figure 4.6c represent the housing recreational facilities, while the purple icon refers to a nearby community mall. From the heat maps, we notice some hot spots that the user has visited while at the home region, but they do not form a stay point. To check whether that specific hotspot is observed from the heat maps, the corresponding location was validated with ground truth from user H. It turns out that user H had only visited the location for a short period of time and stayed for a shorter stay duration than the predefined stay time threshold. Hence, the stay point is not formed due to short duration and can only be observed through heat maps. This demonstrates that the GPS + Wi-Fi stay point detection method is capable of detecting neighbourhood activity to provide in-depth information about the daily trajectory of a user.

Extending beyond the case of a single user, a multi-user neighbourhood study can also be performed. Figure 4.7 shows the neighbourhood activity obtained using the proposed method for three users who reside in the same neighbourhood. The POI visit ground truth for these three users is unknown. Figure 4.7a shows the raw

(a) Raw GPS data    (b) GPS Stay Points by 3 Users    (c) Neighborhood Activity by 3 Users using GPS+WiFi

**FIGURE 4.7**    Comparison of the neighbourhood activity between three users, data collected from 01 August 2020 to 15 March 2021. Note that heatmaps in GPS + Wi-Fi are represented by three different colours (red, green, and blue) to indicate different users' trajectory.

GPS points for the three users. Figure 4.7b shows the GPS stay points for the three users, while Figure 4.7c shows the GPS + Wi-Fi stay points (blue pins) and the home locations (pink pins) for each user. A visual comparison between the figures shows that the inclusion of Wi-Fi data cleans up many inaccuracies from the raw GPS stay points for the three users.

Within Figure 4.7c, the POIs become clearer in areas 1, 5, and 6 as compared to Figure 4.7b. The Wi-Fi information helps in the identification of GPS stay points that belong to the same POI. The heatmap indicated by area 2 shows that the users walk along the river side. This insight is not distinguishable in Figure 4.7b. In addition, while the users' residential POI are found in area 3, there are several other POIs in area 3 as well, including those believed to be void decks directly underneath the user's home. Once again, these POIs are indistinguishable from their home in Figure 4.7b. Finally, a new POI has been identified in area 4.

The knowledge of unexpected POIs visited by multiple users, as well as a geographical representation of commonly visited POI, can help to inform urban planners during the process of planning for layouts of future residential estates and facility areas. With a larger sample size, it may even be possible to infer utilisation rates of specific areas within the estate as well.

### Micro Mobility Study

Beyond examining individual POIs around the neighbourhood, another important aspect of mobility studies is to examine the common routes taken by users between POIs. To extract such insights using both Wi-Fi and GPS data, a data processing pipeline is implemented as shown in Figure 4.8. The timeline extraction is performed on the GPS stay points to obtain the exact moments needed to filter Wi-Fi samples from the Wi-Fi trajectories. Neighbourhood Wi-Fi trajectory data are then clustered together using DBSCAN for all users who live in the same neighbourhood. A Wi-Fi-based clustering process is shown in Algorithm 4.

The aim of the Micro Mobility Study is to identify a similar trajectory path instead of the stay points as identified in the previous two studies. In other words, the

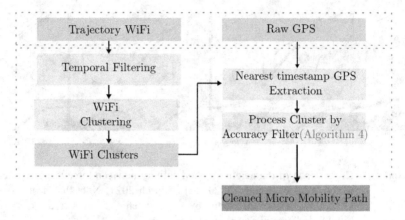

**FIGURE 4.8** Micro mobility analysis data processing pipeline.

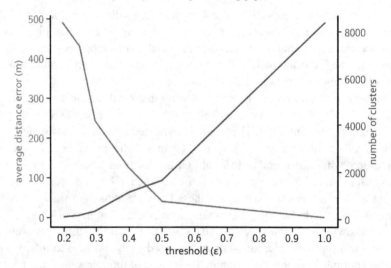

**FIGURE 4.9** Comparison of different threshold values with average distance error ($m$) and number of clusters.

objective is to clear up a messy interpretation of a GPS map as shown in Figure 4.9a into a clearer map as shown in Figure 4.9c. Therefore, the DBSCAN parameters are different in this scenario. The parameter minPts is set to be equal to 1 as it is necessary to include every Wi-Fi scan result into the clustering process, due to the low sampling rate of five minutes that may allow users to travel a substantial distance during that time. After evaluation of the clustering results, a threshold level for cluster formation ($\epsilon$) is selected to provide enough number of clusters to represent the user travel path, which reduces the average distance error in Wi-Fi-based GPS clusters. Moreover, the number of APs observed outdoors is below the low AP level ($A_L$).

Once the clustering is completed, the nearest GPS point for each cluster point's timestamp is extracted. If there is more than one member in a cluster, the cluster representation of the GPS point is determined as follows:

---

**Algorithm 4** The process of extracting WiFi based micro mobility clusters

---

**Input:** Trajectory WiFi ($S_T$) and GPS ($L_T$), $\epsilon$, $minPts$
**Output:** Processed WiFi based clusters ($N_p$)
$N_P = 0$
  Cluster list ($C$) = DBSCAN ($S_T$, $\epsilon$, $minPts$)
  **for** *every index i in C* **do**
    $L_n$ = nearest GPS($t$, $C[i]$, $L_T$)
      **if** *accuracy* $(a) \leq a_L$ **then**
      |   add average $C[i]$ to $N_P$
      **else**
      |   get lowest accuracy, add to $N_P$

      **end if**
  **end for**

---

To define high and low GPS accuracy, an accuracy radius threshold $a_L = 25$ m is used. If more than one member is present in a particular cluster, the average of the nearest GPS points with high accuracy (i.e., accuracy radius $\leq a_L = 25$ m) is calculated and used to represent one Wi-Fi-based cluster with one GPS point. If all the members in a cluster indicate low GPS accuracy (i.e., accuracy radius $> a_L = 25$ m), the GPS point with the lowest accuracy radius value (which means the highest GPS accuracy) is selected, and the rest of the members in the cluster are discarded. Figure 4.9 shows the comparison between the number of clusters, average distance error (in metres) of the cluster points, versus different threshold values for Wi-Fi-based clustering.

Figure 4.9 shows that when the threshold value increases, the number of clusters also increases, and the average distance error decreases. The current objective is to reduce the number of clusters (to obtain a clearer mobility path) and to reduce the average distance error. When $\epsilon = 0.25$, there are 140 clusters with 430.1 m of distance error. In contrast, when $\epsilon = 0.3$, there are 321 clusters with 241.1 m of distance error. Therefore, by considering this trade-off, the value $\epsilon = 0.3$ is selected as the threshold value for Wi-Fi-based clustering. It gives enough clusters to represent a messy GPS micro mobility path into a clearer path while having a reduced average distance error.

For implementation on the data, the mobility pattern of three users is explored. These three users live in the same neighbourhood (i.e., Simei area in Singapore) and work at the same place (i.e., Singapore University of Technology and Design). Most of the time, they commute by walking. The results of the Wi-Fi-based clustering are plotted in Figure 4.10. Figure 4.10a shows the raw GPS for three different users, which consists of GPS data points within the travel duration from individual home to work. Note that each colour denotes a separate user L-Purple, A-Green, and B-Black, and not all the users have the same data amount despite the same timeline, which is from 01 December 2019 to 31 December 2019.

Based on the raw data, data processing is performed as described earlier, and the locations are clustered based on Wi-Fi similarities to preserve significant GPS points. A total of 8,345 raw GPS points were simplified into 140 ($\epsilon = 0.25$) and 321

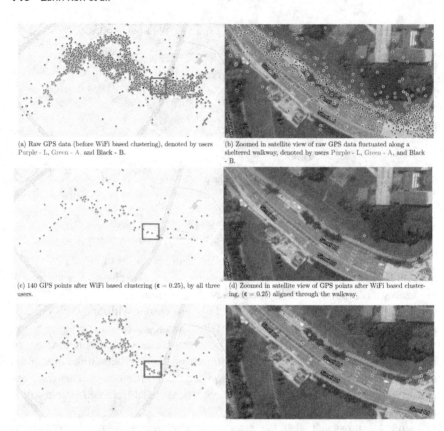

(a) Raw GPS data (before WiFi based clustering), denoted by users Purple - L, Green - A, and Black - B.

(b) Zoomed in satellite view of raw GPS data fluctuated along a sheltered walkway, denoted by users Purple - L, Green - A, and Black - B.

(c) 140 GPS points after WiFi based clustering (ε = 0.25), by all three users.

(d) Zoomed in satellite view of GPS points after WiFi based clustering, (ε = 0.25) aligned through the walkway.

**FIGURE 4.10**   Comparison of before and after Wi-Fi-based clustering for different threshold values; 8,345 points are reduced into 140 points ($\epsilon = 0.25$) and 321 points ($\epsilon = 0.3$).

*Source*: Author.

($\epsilon = 0.3$) clusters, as shown in Figure 4.10c and Figure 4.10e, respectively, based on the Wi-Fi fingerprint clustering method. In other words, each point in Figures 4.10c and 4.10e represents one Wi-Fi-based cluster, which is mapped into the nearest GPS point by timestamp.

Figures 4.10b, d, and f show a zoomed-in satellite view of raw GPS points shown in the red square area in Figures 4.10a, c, and e, respectively. The red square area has a sheltered walkway at the side of the road. By comparing Figures 4.10b, 4.10d, and 4.10f, we can see that when $\varepsilon = 0.3$, the clustered points are aligned through the walkway. Therefore, Wi-Fi-based GPS clustering helps to identify micro mobility patterns of users, which is not possible by only visualising raw GPS data.

The knowledge of these micro mobility patterns allows urban planners to observe common routes taken by users travelling between common POIs. This can provide information such as route choice or utilisation of certain walkways or man-made paths between POIs in a region.

## Case Study 1: Discussion

Within this section, a MCS system was introduced with the aim of understanding three major insights for urban mobility analysis through Wi-Fi fingerprint clustering. GPS location and surrounding Wi-Fi AP data, collected via a smartphone application, were used to identify the indoor POI within a building, obtain neighbourhood activity, and understand micro mobility patterns of the users.

It was demonstrated that, through the fusion of Wi-Fi AP information along with GPS data, it is possible to identify the indoor POI among different users, which is not possible using only GPS location data. Neighbourhood activity analysis is then used to identify the POI where users visit for a short break, while staying at home (e.g., visit a common area in the same building, but a different floor level). The inclusion of Wi-Fi AP information helps to overcome the limitations of using only GPS. These limitations include GPS data being unable to distinguish between different POIs in a multi-level building, as well as inaccuracies from users passing under covered walkways or overhead building structures. The possibility of interpreting the user mobility paths by Wi-Fi clustering-based GPS points for the purpose of identifying the common trajectories was demonstrated. Possible applications of these GPS and Wi-Fi data analysis techniques presented here include the information of residential facility layout planning, route planning, as well as layout planning of different stores within a shopping complex or large buildings.

For the techniques presented in this section to be put into place, it is required to have both the GPS and the Wi-Fi data of each user. There may be cases in which both types of data are not available simultaneously. The following sections will demonstrate other data analysis techniques for insight extraction when only GPS data or only passively collected Wi-Fi data is available.

## Analysis of Individual Users' Mobility Using GPS Trajectory Data

### Case Study 2: Background

From the previous section, it was demonstrated that the use of GPS data, supplemented with the data of surrounding Wi-Fi APs, could provide important insights to be used for Smart Community Design. However, both types of data may not be available at the same time. This section focuses on the scenario where GPS data may be used to identify the mobility patterns of the users on a country-wide scale rather than within a restricted region; hence it may be difficult to collect the corresponding Wi-Fi AP data for such a large coverage.

GPS technology has been used widely for collection of spatio-temporal data on different scales (Van der Spek, Van Schaick, De Bois, & De Haan, 2009). The various fields and applications that have used GPS mobility data include finding efficient routes (Ta, Zhao, & Chai, 2016), understanding the progression of infectious diseases (Hast, Searle, Chaponda, Lupiya, Lubinda, Sikalima, Kobayashi, Shields, Mulenga, Lessler, et al., 2019), and prediction or inference of demographic information of users (Solomon, Bar, Yanai, Shapira, & Rokach, 2018; Wu, Yang, Huang, Wang, Chai, Peng, & Liu, 2019).

In many of the studies, GPS data has been analysed in conjunction with other data, such as demographic data (Siła-Nowicka, Vandrol, Oshan, Long, Demsˇar, & Fotheringham, 2016; Long & Reuschke, 2021), supplementary survey data (Ta, Zhao, & Chai, 2016), or even sound and light data (Marakkalage, Sarica, Lau, Viswanath, Balasubramaniam, Yuen, Yuen, Luo, & Nayak, 2018). With increasing privacy concerns in recent years, the difficulty in obtaining such data for large numbers of volunteers has increased as well. Large volumes of human movement data have also been created without such supplementary data. The goal of this study is to explore ways in which insights can be extracted from GPS data without the need for additional external data.

To this end, previous research such as that done by Zhu, Gonder, and Lin (2017) found that the user's socio-demographic role can be predicted with high accuracy using long-term GPS data. This gives rise to the hypothesis that Working and Non-working users may have different mobility patterns. Additionally, the GPS-enriched trajectory survey dataset of Nahmias-Biran, Han, Bekhor, Zhao, Zegras, and Ben-Akiva (2018) contained several distinct clusters of activity-travel patterns, including distinct temporal patterns of different out-of-work activities as well as different leisure activities. This leads us to examine the mobility patterns of Workdays and Offdays separately.

In this study, a new mobility metric is proposed that excludes the effects of home and work locations and uses the user's home location as a reference point. With the assumption that no external data beyond the GPS data is available, an unsupervised machine learning method – clustering – was selected as it can find groups in data without the need for labels or ground truth. The new mobility metric that was computed for each user is then used in conjunction with other features as an input for the clustering algorithm.

GPS data has been widely used in mobility studies with various applications such as finding significant places (Siła-Nowicka, Vandrol, Oshan, Long, Demsˇar, & Fotheringham, 2016), analysing the impact of employment type (Long & Reuschke, 2021), inferring motion modes (Zheng, Li, Chen, Xie, & Ma, 2008), and investigating neighbourhood activity and micro mobility (Marakkalage, Lau, Zhou, Liu, Yuen, Yow, & Chong, 2021). Statistical properties can be obtained (Alessandretti, Sapiezynski, Lehmann, & Baronchelli, 2017) and anomalous trajectories can even be detected using clustering on GPS trajectories (Wang, Qin, Chen, & Zhao, 2018).

Some deeper analyses can only be done using machine learning techniques such as clustering. Clustering of trajectories can be used to discover dense regions and popular sequential patterns (Cesario, Comito, & Talia, 2017), or to discover groups of common taxi routes (Kumar, Wu, Rajasegarar, Leckie, Krishnaswamy, & Palaniswami, 2018). The DBSCAN (Ester, Kriegel, Sander, Xu, et al., 1996) clustering algorithm can be used to cluster locations of pick-up and drop-off points as a starting point for forming statistical models (Tang, Liu, Wang, & Wang, 2015). Gaussian mixture models (Reynolds, 2009) can be used to cluster groups of people based on their frequency of returning to the same location versus exploring new locations (Amichi, Viana, Crovella, & Loureiro, 2019). To select a good clustering method, the work by Scherrer et al. (Scherrer, Tomko, Ranacher, & Weibel, 2018) serves as a good reference point. Their work presented a rigorous selection process for parameters and clustering algorithms. Out of a total of four clustering algorithms, k-means clustering (Lloyd, 1982) was used in at least the first two combinations in terms of their overall ranking for their objective of clustering users based on the large amount of data gathered from a mobile application without ground truth. As their case study shares similarities to the use case of this study, k-means clustering was selected.

For k-means clustering, the features used as input play an important role. Using mobility metrics is one way to increase the chances of having a meaningful interpretation of clustering results. In past work (Solmaz & Turgut, 2019), mobility metrics can be classified into different types – movement-based, link-based, and network-based. For this study, an emphasis is placed on movement-based and link-based metrics such as visit frequency and mean squared distance.

For movement-based metrics, a commonly used metric is radius of gyration (Gonzalez, Hidalgo, & Barabasi, 2008), which has been used in several works in the past (Pappalardo, Simini, Rinzivillo, Pedreschi, Giannotti, & Barabási, 2015; Pepe, Bajardi, Gauvin, Privitera, Lake, Cattuto, & Tizzoni, 2020; Xu, Belyi, Bojic, & Ratti, 2018). The radius of gyration describes the characteristic distance travelled by a user within a specified time and is calculated as the mean squared distance between the user's visited locations and the computed centre of mass of those locations. Within this chapter, this metric is adapted to focus on non-home and non-work locations and will be described in more detail in "Feature Extraction".

With regard to link-based metrics, the ones mentioned in Solmaz and Turgut (2019) include node density and intercontact time, which are difficult to apply in this current dataset. Thus, Origin-Destination (OD) matrices were considered instead. These have been commonly used in literature (Zhou, Lau, Koh, Yuen, & Ng, 2020; Koh, Zhou, Lau, Yuen, Tuncer, & Chong, 2020) for analysing counts of human flows between locations. Based on the theory that they can be used to describe an individual's probability of motion between different locations, like a Markov chain model, which has high prediction accuracy for trajectories (Lu, Wetter, Bharti, Tatem, & Bengtsson, 2013), a method of feature extraction using the OD matrix is implemented as well.

### System Overview

For this study, timestamped GPS data was collected from a group of volunteers via a user-installed smartphone application. The flowchart of this collected data

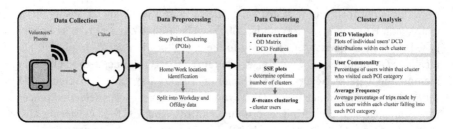

**FIGURE 4.11**    Flowchart depicting the data collection, processing, clustering, and analysis framework proposed by this chapter.

is depicted in Figure 4.11. The criterion for each user's data to be usable was that there had to be at least one month of valid data per user. With this criterion, the data from a total of 73 users was selected for this case study.

The data from each user comes in the form of stay points, where each detected point consists of latitude, longitude, start time, and end time. A validation-based stay point detection algorithm (Lau, Hasala, Kadaba, Thirunavukarasu, Yuen, Yuen, & Nayak, 2017) was used to cluster the individual points at similar coordinates for each user to identify POIs. From this set of POIs, the home and work locations for each user are then detected using frequency and stay duration given the time of day. For example, if a user frequently stayed at a location from late evening to the next morning on most days, it is likely that that location is a Home location for that user. Work locations are detected if a user frequently visits a location and stays there for long durations during the generic working hours from morning to evening on weekdays. This study operates under the assumption of full-time, regular work for working users. If the user does part-time work or has shifts outside of regular working hours, they may not have a detected Work location, and hence they would be categorised as Non-working users for this study.

To investigate the effect of the day type being a Workday as compared to an Offday when the user does not visit their workplace, the POI data was separated into Workday data and Offday data. Workdays are defined as the days when the user was detected at their Work location. If a user is categorised as a Non-working user, all their data is categorised into Offday data.

Next, each POI is labelled according to its proximity to the nearest location. There are ten possible POI labels that were used in this study. If a POI is more than 400 m away from any location with known POI types, it is left unlabelled. The labels and their descriptions are shown in Table 4.6.

The next step was to assign each POI to specific areas called subzones to minimise the impact of the inaccuracy from the GPS signals. Subzones are small sections of planning area delineated by the Urban Redevelopment Agency (URA) for statistical purposes (URA, "Master Plan", 2019). These subzones were used in the extraction of clustering features in the "Proposed Clustering Methodology".

**TABLE 4.6** Labels for the different POI types considered in the dataset

| Label | Description |
| --- | --- |
| Attraction | Places that tourists tend to visit |
| Healthcare | Hospitals, clinics etc. |
| Neighbourhood Centre | Community clubs, hawker centres, markets, etc. |
| Park | Public parks and gardens |
| Places of worship | Temples, mosques, churches, etc. |
| Playground | Playgrounds |
| Recreational | Places that locals tend to visit for leisure |
| Shopping mall | Shopping malls |
| Transportation | Train stations, bus interchanges, etc. |
| Residential | Condominiums, public housing estates, etc. |

*Source*: Authors.

## Proposed Clustering Methodology

The feature extraction and clustering process is described here. For each of the Workday datasets and Offday datasets, the clustering is performed separately.

## Feature Extraction

The feature extraction process used in this study consists of two portions. The first portion of features is adapted from a proposed metric, Daily Characteristic Distance (DCD), while the other is derived from the OD matrix of each user's individual trips.

The proposed metric DCD is derived from a common metric in the literature to measure the mobility of individuals – the radius of gyration. The radius of gyration was proposed by Gonzalez et al. (Gonzalez, Hidalgo, & Barabasi, 2008). Denoted as $r_g^a(t)$ the radius of gyration of a user $a$ from the start of their dataset up to a certain time $t$ is expressed by Equation 11.

$$r_g^a(t) = \sqrt{\frac{1}{n_c^a(t)} \sum_t^{n_c^a} \left( \vec{r}_i^a - \vec{r}_{cm}^a \right)^2} \tag{11}$$

In Equation 11, $\vec{r}_i^a$ represents the $i = 1, \ldots, n_c^a(t)$ positions recorded for user $a$ and $\vec{r}_{cm}^a = (1/n_c^a(t)) \sum_{i=1}^{n_c^a} \vec{r}_i^a$ is the centre of mass of the trajectory.

As the duration of data collected can vary between users, the time dependency should be removed to ensure a fair comparison. The value of $n_c^a(t)$ is replaced with the total number of locations $N^a$ visited by user $a$. The simplified equation is as shown in Equation 12.

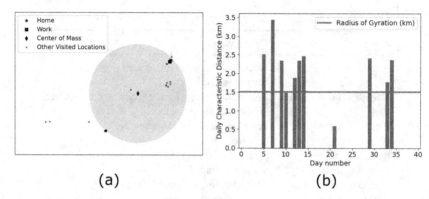

(a)                                                    (b)

FIGURE 4.12    (a) Illustration of radius of gyration. (b) Comparison between radius of gyration (single value, red line) and proposed DCD (set of values, bar plot) over the same time period.

*Source*: Authors.

$$r_g^a = \sqrt{\frac{1}{N^a} \sum_{i=1}^{N} \left( \vec{r}_i^a - \vec{r}_{cm}^a \right)^2} \qquad (12)$$

An illustration for this time-simplified radius of gyration metric is shown in Figure 4.12a for a user with a 40-day dataset. The centre of mass is observed to lie between the Home and Work location, which is in line with expectations as those two locations are the ones visited most frequently. The shaded circle is centred on the centre of mass and has a radius equal to the computed radius of gyration.

To add meaning to this metric, further adaptations can be made. As the Home location is part of the dataset, it can be used as the reference point for distance calculations instead of the computed centre of mass of all the user's visited locations. This allows for the inference of how far the user generally travels from their Home location, which may have more physical meaning than the computed centre of mass of their trajectories. Next, the radius of gyration currently produces a single value for each user regardless of the actual duration of their dataset, as we consider the whole duration. To obtain more information about how the user travels differently on different days, the user's dataset is broken down into individual days and each day will have one value computed for that specific day's data. With this, a distribution of the user's daily distances travelled can be obtained. Finally, as the focus of this work is on the non-Home and non-work locations, the influences of the Home and Work locations on the computation are manually removed. To achieve this, the distances of Home and Work locations are manually set to zero, and the counts of Home and Work visits are subtracted from the total count of visited locations $N^a$. With the addition of these adaptations, the new metric is named

Daily Characteristic Distance (DCD). The formula is to calculate DCD for a given day $d$ is shown in Equation 13.

$$DCD_d = \sqrt{\frac{1}{n_d}\sum_{i=1}^{n_d} f_{id} \times \left(\vec{r}_{id} - \vec{r}_{home}\right)^2} \tag{13}$$

In Equation 13, $n_d$ represents the number of unique POIs that the user travelled to on that day, $f_{id}$ is the number of times the user travelled to location $i$ where $i = 1,\ldots, n_d$ on that day, $r_{home}$ is given by the mean coordinates of the Home location of the user. Figure 4.12 illustrates the difference between the radius of gyration (one value per user) and DCD (a set of values per user). The days without bars have a value of zero, indicating that on those days the user travelled directly between Home and Work without visiting any other location, or that the user simply stayed at Home that day.

Figure 4.13 shows the obtained DCD distributions of all users within the dataset. Figure 26a depicts the Workday distributions of Working users, while Figure 26b shows the Offday data of all users. The Offday data includes the Offday data of Working users and all the data of Non-working users.

In Figure 4.13(a), the distributions are sorted in ascending order of Home-Work distance of each user. It may be observed that there is a likely relationship between the Home-Work distance of the users and the location of the peaks of their DCD distributions. The Pearson's $R$-value between each user's Home-Work distance and the median of their DCD distribution was computed to be a moderately high $R$-value of 0.746, with a $p$-value of $2.90 \times 10^5$.

For the Offday data, the distributions were sorted in ascending order of their median distance and coloured according to whether the user is a Working or Non-working user. From Figure 4.13(b), it can be observed that there is a higher

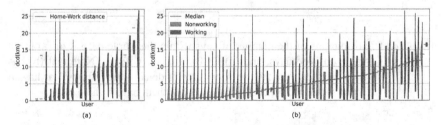

(a)    (b)

**FIGURE 4.13**    Violinplots illustrating the DCD distributions of users on (a) Workdays, consisting of only Working users, and (b) Offdays, consisting of both Working users on Offdays and Non-working users on all days. (a) shows a moderately high correlation between DCD peaks and Home-Work distance of each user, while (b) shows a higher density of Working users with higher median DCD.

*Source*: Authors.

concentration of Working users (blue) at the side with higher median DCD. This may indicate that Working users tend to visit locations at further distances from their homes as compared to Non-working users.

After computing DCD for every user, the DCD values are then separated into Workday and Offday DCD values. A histogram is then plotted for each set of values to find suitable thresholds for assigning feature values to each DCD. The histograms are plotted in Figure 4.14a and 4.14b for Workday DCD and Offday DCD, respectively. From these, histograms, 0–5 km, 5–15 km, and >15 km, can be obtained. One extra range of 0 km is included to denote Home and/or Work only. Thus, a total of four distance ranges for DCD are used.

Next, for each user's data, the percentage of DCD values that fall within each of the determined thresholds is calculated accordingly. This will give four features for each type of data that adds up to 1.0. An example of the features for one user, User 2, is shown in Table 4.7.

The second part of the features is extracted from the OD matrices from the users' data. Since the POIs visited by different users are different, it is not feasible to use all the different POIs to create OD matrices. Instead, the distances travelled by users to each POI from either their Home or their Work location can be measured.

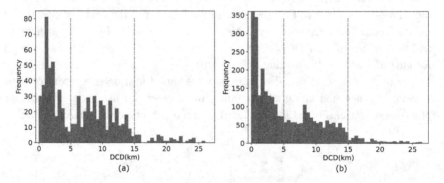

(a)  (b)

**FIGURE 4.14**  Histograms of the number of days within the whole dataset of (a) DCD value on Workdays, and (b) DCD value on Offdays. Note that (b) has been cropped vertically to show greater detail – the leftmost bar has an actual value of 3,312, of which 3,137 of them have a value of 0.

*Source*: Authors.

**TABLE 4.7**  Example of the four DCD features for Workday and Offday data

| Day Type | Home/Work | 0–5 km | 5–15 km | >15 km |
| --- | --- | --- | --- | --- |
| Workday | 0.60 | 0.23 | 0.15 | 0.02 |
| Offday | 0.38 | 0.16 | 0.41 | 0.05 |

*Source*: Authors.

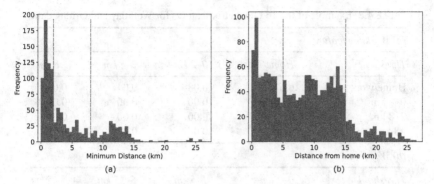

**FIGURE 4.15** Histograms of the number of POIs visited over the whole dataset of (a) minimum distance between Home and Work to that location on Workdays, and (b) distance from home to that location on Offdays.

*Source*: Authors.

For Workdays, the minimum of the distances between Home to POI and Work to POI is taken as the distance to be considered, as some users may visit a POI close to their Work location as it is more convenient there. To say that the user has truly gone out of their way to visit a certain POI, both the distances between Home to POI and Work to POI need to be larger. For Offdays, there is no such common location with meaning for all the users; hence the distance from Home to POI can be used. After extracting these distances separately from the Workday and Offday data, the corresponding histograms are plotted as shown in Figure 4.15a and 4.15b. The distance ranges for Workdays are 0–2 km, 2–8 km, and >8 km, with an additional 0 km category to denote trips between Home and Work. For Offdays, the ranges for trip distances are 0–1 km, 1–5 km, 5–15 km, and >15 km.

After getting these thresholds, the trips made by each user are now categorised based on these thresholds. Taking an example of a user with Workday data, a trip consists of going from POI A to POI B, where threshold A is on the row of the matrix and threshold B is on the column of the matrix. If A is located within 0–2 km and B is located within 2–8 km, the count corresponding to the "0–2 km" row and the "2–8 km" column will be increased by 1, and so on. One thing to note is that trips occurring on different calendar dates (i.e., from the last POI on one day to the first POI the next morning) are not counted. Trips that occur within the same subzone (e.g., Home to Home) are also not counted. After the trips are all counted for a user, the matrix is normalised by the total number of trips counted for that user, such that all 16 elements of this matrix add up to 1.0. This is to make the data comparable between different users. An example of the resulting matrices using the Workday and Offday thresholds can be seen in Table 4.8.

These 16 O-D features are combined with the four features from the above DCD calculations to form the 20 features. These 20 features describe a user's mobility pattern for each type of day by showing the proportion of trips made by the user to and from each distance threshold.

**TABLE 4.8** Example of the 16 OD matrix features for Workday and Offday data

*(a) Workday features*

| Threshold | Home/Work | 0–2 km | 2–8 km | >8 km |
|---|---|---|---|---|
| Home/work | 0.67 | 0.08 | 0.04 | 0.04 |
| 0–2 km | 0.08 | 0.00 | 0.00 | 0.00 |
| 2–8 km | 0.04 | 0.00 | 0.00 | 0.00 |
| >8 km | 0.03 | 0.00 | 0.00 | 0.01 |

*(b) Way features*

| Threshold | 0–1 km | 1–5 km | 5–15 km | >15 km |
|---|---|---|---|---|
| 0–1 km | 0.00 | 0.08 | 0.23 | 0.03 |
| 1–5 km | 0.08 | 0.03 | 0.00 | 0.00 |
| 5–15 km | 0.19 | 0.04 | 0.29 | 0.00 |
| >15 km | 0.02 | 0.00 | 0.01 | 0.00 |

*Source*: Author.

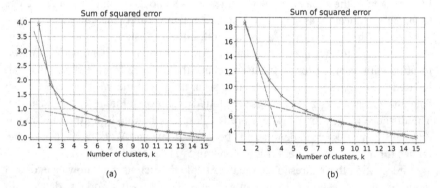

(a)  (b)

**FIGURE 4.16**  SSE plots used to derive (a) the optimal number of clusters for Workdays and (b) the optimal number of clusters for Offdays. Both plots indicate 3 as a suitable value for *k*, the number of clusters.

*Source*: Authors.

### Clustering Process

The selected clustering method, *k*-means clustering (Lloyd, 1982), requires an input parameter *k*, which is the number of clusters. To determine an optimal value for this parameter, a sum-of-squared errors (SSE) plot is used. The SSE plot measures the sum of all squared errors from the clustered points to their respective cluster centres after clustering using each value of *k*. A good value for *k* would be located just before the decrease in SSE becomes less than proportionate to the increase in *k*. The SSE plots for this dataset can be seen in Figure 4.16, where Figure 4.16a shows the plot using the data from workdays, while Figure 4.16b shows the plot using the data from Offdays. From both SSE plots, the "elbow" of the plot indicates that a good value of $k$ to use would be $k = 3$.

## Centroid Values

After clustering, the first aspect of the cluster results to examine is the cluster centres, or centroids, as they can offer some insight into the distinguishing factors of each cluster. The values of each feature in the centroids are the mathematical average values of each feature within each cluster. The Workday and Offday datasets were clustered separately, and their cluster centres are plotted in Figures 4.17 and 4.18, respectively. The Workday clusters are referred to as W1, W2, and W3, respectively, and the Offday clusters are O1, O2, and O3.

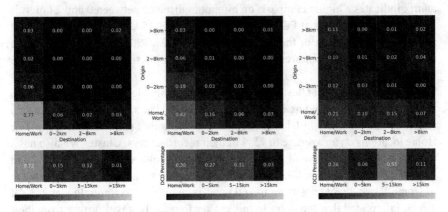

**FIGURE 4.17** Centroid values of the three clusters obtained from clustering Workday data.

*Source*: Authors.

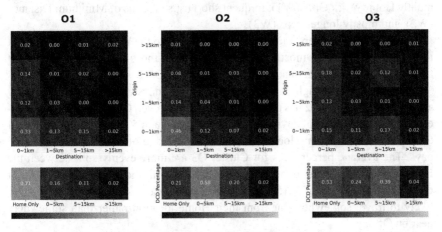

**FIGURE 4.18** Centroid values of the three clusters obtained from clustering Offday data.

*Source*: Authors.

For the Workday clusters, one apparent distinguishing factor between the three clusters is the proportion of "Home/Work" to "Home/Work" trips compared to trips to POI at other distance thresholds. Cluster W1 has the highest percentage of trips directly between Home and Work, as well as the highest average percentage of days spent only at Home or Work. The other two clusters W2 and W3 are in descending order of percentage of trips directly between Home and Work.

The users in Cluster W1 have a large majority – on average 72% – of Workdays when they do not visit any non-Home, non-work locations. The average percentage of their days spent with DCD at each distance threshold decreases with increasing distance. In contrast, for Cluster W2, the DCD features are roughly evenly spread across the first three distance thresholds, with a higher value of DCD being within 5–15 km as compared to 0–5 km. The percentage of trips from the OD matrix indicates a higher emphasis on minimum distance between 0 and 2 km. It is likely that there may exist POI locations which are 5–15 km from their Home but are also within 0–2 km of their workplace, which leads to a focus on trips in the 0–2 km range for OD features while having more emphasis on the 5–15 km range for DCD features.

The DCD features of Cluster W3 show a remarkably high average value of 55% in the 5–15 km threshold as compared to 12% and 31% in the other two clusters, which means the users in Cluster W3 travel to POIs that are 5–15 km from their Home locations on an average of 55% of their recorded days. Cluster W3 also has a much lower average value of 8% in the 0–5 km threshold, as compared to 15% and 27% in the other clusters. Viewing this in conjunction with the average percentage of Home/Work direct trips from the OD matrix, which is also quite low at 21%, it can be interpreted that the users in this cluster frequently travel further from their Home and Work locations. They also make a larger number of direct trips between POIs that are far from their homes. In summary, the clusters can be described as mainly Home/Work Only (W1), frequent short trips in terms of Minimum Distance (W2), and mostly longer trips (W3).

For the Offday clusters, they show similar trends to the Workday clusters in that there are those that stay mostly at Home Only (O1), those that make mostly short trips (O2), and those that make mostly longer trips (O3).

Cluster O1 has the highest average percentage of days with DCD at Home only at 71%, while Cluster O2 has the lowest at 21%. However, Cluster O2 has the highest average percentage of days with DCD between 0 and 5 km, meaning they went to at least one other non-Home location on an average of 58% of their recorded days. The average percentages for Cluster O3 are more evenly split between the Home Only and the first two distance categories, with the highest being 39% of days with DCD values of 5–15 km. This indicates that users in this cluster are more likely to travel somewhere far from their Home location rather than somewhere near on Offday.

When observing each cluster's corresponding OD matrix, Cluster O2 has the highest average percentage of trips within 0–1 km at 46% compared to 32% for

Cluster O1. Additionally, the percentages of trips going between the 0–1 km threshold and further thresholds in the bottom row is higher in Cluster O1 than in Cluster O2. It is possible that although the users in O1 stay at Home only for more days than those in O2, they tend to travel further when they do go out, as compared to those in O2 who could go out on more days but stay within 0–1 km for most of their trips. The users in Cluster O3 seem to have more of a balance between staying at home and going out to near or further places. Quite a high percentage of their trips are also made within the 5–15 km range.

### DCD Violinplots

For further analysis of each cluster, the violinplots of DCD for each user in each cluster are plotted to show the distribution of DCD values for each user for their entire recorded duration. The violinplots are sorted in ascending order of their Home-Work distance. These violinplots do not include the percentage of days spent only between Work and Home, as the focus of this study lies in examining the POIs that are not Home and not Work. The violinplots for the Workday clusters are shown in Figure 4.19 while those for Offday clusters are shown in Figure 4.20.

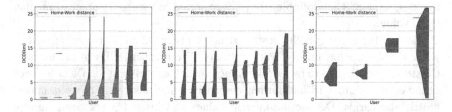

**FIGURE 4.19**   Violinplots illustrating each user's DCD distribution within each cluster on Workdays.

*Source*: Authors.

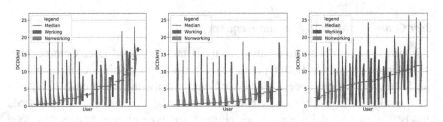

**FIGURE 4.20**   Violinplots illustrating each user's DCD distribution within each cluster on Offdays.

*Source*: Authors.

From Figure 4.19, the yellow highlighted portion shows that most of the users in Cluster W1 have a low Home-Work distance, which is below 5 km. This may be a factor in these users having the highest percentage of direct trips between Home and Work out of the three clusters on Workdays, as described by their cluster centroid values. The users in Cluster W2 have Home-Work distances in the middle range, and usually the peaks of their DCD distributions are located close to the Home-Work distances. This is also reflected in their OD matrix, in which this cluster has the highest percentage of trips within 0–2 km of either their Home or their Work location out of the three clusters. For Cluster W3, two out of the four users have a large Home-Work distance of over 20 km. Three out of the four users have DCD peaks near their Home-Work distance. However, this is not reflected in the centroid OD matrix. A possible reason could be because they travel to other places that are the same distance from their Home as well as their Work location. These DCD plots correspond to the DCD features for Cluster W3, as the bulk of their DCD distributions are located within the 5–15 km range.

The violinplots representing the DCD distribution of each user within each Offday cluster have been plotted in Figure 4.20. Working and Non-working users have been coloured differently to distinguish them from each other. The Home Only days are not reflected on this plot as the focus lies on days in which the users go out. The violinplots have been sorted in ascending order of median DCD value.

Most of the users in both Cluster O1 and Cluster O2 are Non-working users, while Cluster O3 has mostly Working users. The median DCDs of the users in Cluster O2 are limited to the 0–5 km range, which agrees with the DCD features observed in Figure 4.18 and further emphasises that this group of users makes mostly short trips. Despite the median values of Cluster O3 not being strictly higher than those in O2, the bulk of the DCD distributions for Cluster O3 lies above 5 km, which is the distance threshold for longer trips in this case. It can also be observed that the Working users in all three clusters tend to be towards the higher end. This may hint that Working users tend to make trips over longer distances.

### User Commonality and Average Frequency

This next portion covers two cluster analysis metrics called User Commonality and Average Frequency. Both types of analysis use the same distance thresholds as the feature extraction process. For User Commonality, the intention is to answer the question: is there any specific combination of POI type and distance threshold that is favoured by the users in each cluster? The type of each POI is described by the label assigned to it, and the list of labels can be found in "Data Collection and Preprocessing". The computation is as follows. For each combination of distance threshold $j$ and label $k$, the User Commonality value $u_{jk}$ is calculated using

Equation 14, where $n_{jk}$ is the number of users within the cluster who visited a POI at distance threshold $j$ with label $k$, and $u_{jk}$ is the total number of users in that cluster.

$$u_{jk} = \frac{n_{jk}}{n_c} \tag{14}$$

The computed User Commonality values are then combined into a table and plotted into a heatmap for visualisation.

For Average Frequency, the computation process covers two steps. First, the number of each user $i$'s total trips at each combination of distance threshold $j$ with label $k$ is extracted as $P_{ijk}$ and normalised by the total number of labelled POIs visited by user $i$, denoted as $p_i$. This is shown by Equation 15, where $P_{ijk}$ represents the fractional frequency of user $i$ visiting a POI at distance threshold $j$ with label $k$.

$$P_{ijk} = \frac{P_{ijk}}{p_i} \tag{15}$$

Next, the average value of all users' fractional frequency within a cluster is calculated for each distance threshold $j$ and label $k$. This value $f_{jk}$ is referred to as the Average Frequency value in Equation 16, where $n_c$ represents the total number of users within the cluster that the user belongs to.

$$f_{jk} = \frac{\sum_{i=1}^{n_c} P_{ijk}}{n_c} \tag{16}$$

All the Average Frequency values $f_{jk}$ for each distance threshold $j$ with label $k$ are combined into a table and plotted into a similar heatmap as User Commonality to facilitate a side-by-side comparison and interpretation. The User Commonality and Average Frequency heatmaps for the Workday clusters (W1 to W3) are shown in Figure 4.21, while those for the Offday clusters (O1 to O3) are shown in Figure 4.22. The User Commonality heatmaps are labelled in the figures as (a) while the Average Frequency heatmaps are labelled in the figures as (b).

With regard to User Commonality plotted in Figure 4.21(a), Clusters W1 and W2 do not contain a single-distance threshold and POI label combination that is visited by 100% of the users. In contrast, Cluster W3 has 100% of its users visit POIs labelled as "Shopping Mall" that are located at a minimum distance of >8 km from both their Work and their Home locations. However, this may not be a unique distinguishing factor of Cluster W3 as quite a high percentage of users in the other two clusters visit this distance threshold/POI label combination as well. Other commonly visited combinations that appear in all three clusters are the 0–2 km threshold for the POI labels of Neighbourhood Centre, Shopping Mall, and Residential. The distinguishing features of each Workday cluster can be summarised as follows. Cluster W1 has a visible percentage of users who visit Recreational POIs at >8 km minimum distance as compared to the other two clusters.

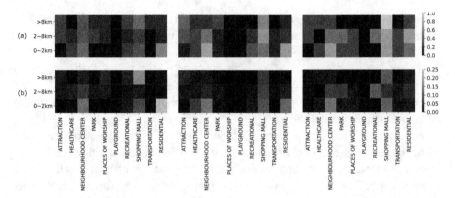

**FIGURE 4.21** Heatmaps for each of the three Workday clusters showing (a) User Commonality and (b) Average Frequency. The colourmap scales for (b) are narrowed to 0.25 to better show the contrast between the different squares.

*Source*: Author.

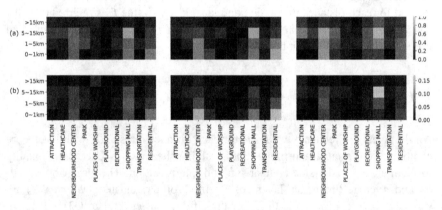

**FIGURE 4.22** Heatmaps for each of the three Offday clusters showing (a) User Commonality and (b) Average Frequency. The colourmap scales for (b) are narrowed to 0.17 to better show the contrast between the different squares.

*Source*: Authors.

More of the users in Cluster W2 visit Attractions at a minimum distance of larger than 8 km as compared to the other clusters. Cluster W3 has a much higher percentage of its users that go to Residential POIs at the 2–8 km threshold as compared to the other two clusters, which have more users visiting Residential POIs at the 0–2 km threshold.

For the Average Frequency between the Workday clusters, the users in all three clusters have a high average frequency of visits to Neighbourhood Centre POIs at 0–2 km. Cluster W1 has the highest Average Frequency at Residential POIs within the 0–2 km threshold and Shopping Mall POIs at the >8 km threshold. Cluster W2 has a visible frequency at Healthcare POIs at the 0–2 km distance threshold. This is not observed in the other two clusters and could be a distinguishing factor. For Cluster W3, there is a visible frequency at the Park and Recreational POIs within the 2–8 km distance thresholds, which is not observed in the other two clusters.

Comparing both parts (a) and (b) of Figure 4.21, it can be observed that some combinations are visited by more users in each cluster but may not be visited very frequently as a proportion of each individual user's visits. Examples of this include the Attraction POIs at the 2–8 km and >8 km distance thresholds, Park POIs for W2, and Places of Worship for W3.

For the Offday Clusters O1 to O3, Figure 4.22 shows the User Commonality and Average Frequency heatmaps derived from the same process as those in Figure 4.21. At first glance, the User Commonality figures of the three Clusters O1, O2, and O3 appear much more similar to each other as compared to the Workday clusters of W1, W2, and W3. The three main POI labels visited most commonly by users in all three clusters are Neighbourhood Centre, Shopping Mall, and Residential. Some combinations that are highlighted in only one cluster are Places of Worship at the 5–15 km threshold for Cluster O2 and Recreational at the 1–5 km and 5–15 km thresholds for Cluster O3.

From Figure 4.22(b), the labels Neighbourhood Centre, Shopping Mall, and Residential are the most prominent in terms of Average Frequency for all three clusters as well. For Shopping Malls, the distance threshold with the highest frequency differs between each cluster. For Cluster O1, the range with the highest Average Frequency is 1–5 km, while that for Cluster O2 is 0–1 km, and for Cluster O3, it is 5–15 km. For Cluster O2 specifically, this highest frequency distance threshold is the same for the three labels of Neighbourhood Centre, Shopping Mall, and Residential. It may be inferred that the frequent short trips made by users in Cluster O2 (observed from Figure 4.18) are for the purpose of visiting these three types of locations. For Cluster O3, the average frequency over all the distance thresholds and POI combinations is prominently concentrated at Shopping Malls in the 5–15 km distance threshold, and the frequency for Residential areas is much lower than for the other two clusters. It can be inferred that users in this cluster frequently make the effort to visit shopping malls that are further away from their home locations as compared to users in the other two clusters, even on Offdays where it would not be convenient to visit before or after work.

## Case Study 2: Discussion

In this section, the differences between the GPS trajectory patterns of Workday and Offday data as well as the differences between the Working and Non-working users were investigated. In the process, the proposed mobility metric DCD was used to zoom in on the locations visited by each user outside of their Home and

Work locations (if applicable). Features extracted using DCD and OD matrices were used as input for a clustering algorithm, which was able to cluster users based on how far they usually travelled and how often they travelled to each distance threshold.

From the feature extraction and preliminary data visualisation of DCD distributions, it was observed that Working users' median DCD on Workdays is highly correlated to the distance between their Home and Work locations, indicating that the daily distances travelled by Working users to other places of interest are quite balanced around the distance between their Home and Work locations. Working users generally have a higher median DCD on Offdays as compared to Non-working users, which may imply that Working users tend to travel further from their Home location even on Offdays without having work as an added reason to travel.

From the clustering results, it was found that users could be grouped into three types for both Workdays and Offdays, based on their DCD distributions and the frequency of trips between each combination of distance thresholds. These three types of users include those that mainly travel only between their Home and Work locations (if applicable), those that make frequent trips to locations within a short distance from their Home and Work locations (if applicable), and those that make trips to places further away that may not be as frequent.

For further cluster analysis, the two metrics User Commonality and Average Frequency were introduced. These help to paint a picture of the types of locations and distance thresholds that are visited by more of the users of the same cluster as well as those visited more frequently on average by the users within each cluster. Three main POI types that are favoured regardless of cluster are Neighbourhood Centres, Shopping Malls, and Residential areas. Some differences between clusters are the presence or absence of visits to other POI types such as Attractions and Parks. The insights extracted from knowing the favoured POI types and distances of each cluster can be used for urban planning of different types of facilities at different distances from each other with target audiences from each cluster.

For future work, some things may be altered to better suit the needs of each individual study. For example, if there are any specific distance thresholds or POI types that may be meaningful to the objectives of the study, those may be purposefully implemented within the data preprocessing and feature extraction process instead of using a fully data-based approach like the one presented in this chapter. The clustering and data analysis portion can still be implemented in the same manner according to the selected features.

## Conclusion

In this chapter, to support Smart Community Design, various methods of data analysis were explored to understand urban human mobility within the neighbourhood and beyond, with a focus on data analysis utilising two sources of data gathered from smart mobile devices – GPS data and passively sensed Wi-Fi data.

Under the analysis of urban mobility via clustering of Wi-Fi signals in case study 1, a method of Wi-Fi fingerprint clustering for urban mobility was introduced. The fusion of Wi-Fi AP information along with GPS data was demonstrated to enable the identification of the indoor POIs among different users, which are otherwise impossible to identify using only GPS location data. Neighbourhood activity analysis is then used to identify the POI where users visit for a short time while staying at home, with variations in vertical location within a single block. The inclusion of Wi-Fi AP information allowed only the limitations of using GPS to be overcome, which include being unable to differentiate between vertical locations. The insights gathered from this section can be used to inform planning of POIs in large buildings like shopping complexes or analysis of common routes taken with sheltered walkways and so on.

For case study 2 when GPS data for individual users is available, the analysis of individual users' mobility using GPS trajectory data presented a new mobility metric DCD for use with GPS data linked to individuals. The correlation of how DCD changes with the status (Working and Non-working) of the users, as well as the specific day type (Workdays and Offdays), was presented. DCD and OD matrix features were extracted as a representation of each user, which were then clustered. Afterwards, a further analysis of the cluster results was performed using two other proposed analysis metrics – User Commonality and Average Frequency. The results of this section can profile different types of people and their willingness to travel longer distances for different types of POIs and allow urban planners to plan the locations of facilities on a city-level scale.

## Bibliography

Alessandretti, L., Sapiezynski, P., Lehmann, S., & Baronchelli, A. (2017). Multi-scale spatio-temporal analysis of human mobility. *PLoS One, 12*(2), e0171686.

Alzantot, M., & Youssef, M. (2012). Crowdinside: Automatic construction of indoor floorplans. In *Proceedings of the 20th International Conference on Advances in Geographic Information Systems* (pp. 99–108). New York: ACM.

Amichi, L., Viana, A. C., Crovella, M., & Loureiro, A. F. (2019). Mobility profiling: Identifying scouters in the crowd. In *Proceedings of the 15th International Conference on Emerging Networking EXperiments and Technologies* (pp. 9–11). New York: ACM.

Blondel, V. D., Guillaume, J. L., Lambiotte, R., & Lefebvre, E. (2008). Fast unfolding of communities in large networks. *Journal of Statistical Mechanics: Theory and Experiment, 2008*(10), P10008

Cesario, E., Comito, C., & Talia, D. (2017). An approach for the discovery and validation of urban mobility patterns. *Pervasive and Mobile Computing, 42*, 77–92.

Ester, M., Kriegel, H. P., Sander, J., Xu, X., et al. (1996). A density-based algorithm for discovering clusters in large spatial databases with noise. *Kdd, 96* (34), 226–231.

Farkas, K., Feher, G., Benczur, A., & Sidlo, C. (2015). Crowdsensing based public transport information service in smart cities. *IEEE Communications Magazine, 53*(8), 158–165.

Gamanayake, C. M., Jayasinghe, L. A., Ng, B., & Yuen, C. (2020). Cluster pruning: An efficient filter pruning method for edge AI vision applications. *IEEE Journal of Selected Topics in Signal Processing, 14*(4), 802–816.

Ganti, R. K., Ye, F., & Lei, H. (2011). Mobile crowdsensing: Current state and future challenges. *IEEE Communications Magazine, 49*(11), 32–39.

Gonzalez, M. C., Hidalgo, C. A., & Barabasi, A.-L. (2008). Understanding individual human mobility patterns. *Nature, 453*(7196), 779.

Google. (2018). *Wi-Fi scanning*. Retrieved from https://goo.gl/RqxNk2

Hast, M., Searle, K. M., Chaponda, M., Lupiya, J., Lubinda, J., Sikalima, J., Kobayashi, T., Shields, T., Mulenga, M., Lessler, J., et al. (2019). The use of GPS data loggers to describe the impact of spatio-temporal movement patterns on malaria control in a high-transmission area of northern Zambia. *International Journal of Health Geographics, 18*(1), 1–18.

Helgason, Ó., Kouyoumdjieva, S. T., & Karlsson, G. (2013). Opportunistic communication and human mobility. *IEEE Transactions on Mobile Computing, 13*(7), 1597–1610.

Hoteit, S., Secci, S., Sobolevsky, S., Ratti, C., & Pujolle, G. (2014). Estimating human trajectories and hotspots through mobile phone data. *Computer Networks, 64*, 296–307.

Hu, X., Li, X., Ngai, E., Leung, V., & Kruchten, P. (2014). Multidimensional context-aware social network architecture for mobile crowdsensing. *IEEE Communications Magazine, 52*(6), 78–87.

Kang, C., Sobolevsky, S., Liu, Y., & Ratti, C. (2013). Exploring human movements in Singapore: A comparative analysis based on mobile phone and taxicab usages. In *Proceedings of the 2nd ACM SIGKDD International Workshop on Urban Computing* (p. 1). New York: ACM.

Koh, Z., Zhou, Y., Lau, B. P. L., Yuen, C., Tuncer, B., & Chong, K. H. (2020). Multiple-perspective clustering of passive Wi-Fi sensing trajectory data. *IEEE Transactions on Big Data, 8*(5), 1312–1325.

Kumar, D., Wu, H., Rajasegarar, S., Leckie, C., Krishnaswamy, S., & Palaniswami, M. (2018). Fast and scalable big data trajectory clustering for understanding urban mobility. *IEEE Transactions on Intelligent Transportation Systems, 19*(11), 3709–3722.

Lau, B. P. L., Hasala, M. S., Kadaba, V. S., Thirunavukarasu, B., Yuen, C., Yuen, B., & Nayak, R. (2017). Extracting point of interest and classifying environment for low sampling crowd sensing smartphone sensor data. In *2017 IEEE International Conference on Pervasive Computing and Communications Workshops (PerCom Workshops)* (pp. 201–206). Kona, HI, USA: IEEE.

Lau, B. P. L., Marakkalage, S. H., Zhou, Y., Hassan, N. U., Yuen, C., Zhang, M., & Tan, U.-X. (2019). A survey of data fusion in smart city applications. *Information Fusion, 52*, 357–374.

Leonardi, C., Cappellotto, A., Caraviello, M., Lepri, B., & Antonelli, F. (2014). Secondnose: An air quality mobile crowdsensing system. In *Proceedings of the 8th Nordic Conference on Human-Computer Interaction: Fun, Fast, Foundational* (pp. 1051–1054). New York: ACM.

Li, Q., Zheng, Y., Xie, X., Chen, Y., Liu, W., & Ma, W.-Y. (2008). Mining user similarity based on location history. In *Proceedings of the 16th ACM SIGSPATIAL International Conference on Advances in Geographic Information Systems* (p. 34). New York: ACM.

Liu, R., Marakkalage, S. H., Padmal, M., Shaganan, T., Yuen, C., Guan, Y. L., & Tan, U.-X. (2019a). Crowd-sensing simultaneous localization and radio fingerprint mapping based on probabilistic similarity models. In *Proceedings of the ION 2019 Pacific PNT Meeting, Honolulu, Hawaii, April 2019* (pp. 73–83). ION.

Liu, R., Marakkalage, S. H., Padmal, M., Shaganan, T., Yuen, C., Guan, Y. L., & Tan, U.-X. (2019b). Collaborative slam based on Wi-Fi fingerprint similarity and motion information. *IEEE Internet of Things Journal, 7*(3), 1826–1840.

Liu, R., Yuen, C., Do, T., & Tan, U.-X. (2017). Fusing similarity-based sequence and dead reckoning for indoor positioning without training. *IEEE Sensors Journal, 17*(13), 4197–4207.

Lloyd, S. (1982). Least squares quantization in PCM. *IEEE Transactions on Information Theory, 28*(2), 129–137.

Long, J., & Reuschke, D. (2021). Daily mobility patterns of small business owners and homeworkers in post-industrial cities. *Computers, Environment and Urban Systems, 85*, 101564.

Lou, Y., Zhang, C., Zheng, Y., Xie, X., Wang, W., & Huang, Y. (2009). Map-matching for low-sampling-rate GPS trajectories. In *Proceedings of the 17th ACM SIGSPATIAL International Conference on Advances in Geographic Information Systems* (pp. 352–361). New York: ACM.

Lu, X., Wetter, E., Bharti, N., Tatem, A. J., & Bengtsson, L. (2013). Approaching the limit of predictability in human mobility. *Scientific Reports, 3*(1), 1–9.

Marakkalage, S. H., Lau, B. P. L., Viswanath, S. K., Yuen, C., & Yuen, B. (2019). Real-time data analysis using a smartphone mobile application. In B. Yuen (Ed.), *Ageing and the built environment in Singapore* (pp. 221–240). Cham, Switzerland: Springer.

Marakkalage, S. H., Lau, B. P. L., Zhou, Y., Liu, R., Yuen, C., Yow, W. Q., & Chong, K. H. (2021). Wi-Fi fingerprint clustering for urban mobility analysis. *IEEE Access, 9*, 69527–69538.

Marakkalage, S. H., Liu, R., Viswanath, S. K., & Yuen, C. (2019). Identifying indoor points of interest via mobile crowdsensing: An experimental study. In *2019 IEEE VTS Asia Pacific Wireless Communications Symposium (APWCS)* (pp. 1–5). Singapore: IEEE.

Marakkalage, S. H., Sarica, S., Lau, B. P. L., Viswanath, S. K., Balasubramaniam, T., Yuen, C., Yuen, B., Luo, J., & Nayak, R. (2018). Understanding the lifestyle of older population: Mobile crowdsensing approach. *IEEE Transactions on Computational Social Systems, 6*(1), 81–95.

Nahmias-Biran, B.-H., Han, Y., Bekhor, S., Zhao, F., Zegras, C., & Ben-Akiva, M. (2018). Enriching activity-based models using smartphone-based travel surveys. *Transportation Research Record, 2672*(42), 280–291.

Pappalardo, L., Simini, F., Rinzivillo, S., Pedreschi, D., Giannotti, F., & Barabási, A.-L. (2015). Returners and explorers' dichotomy in human mobility. *Nature Communications, 6*(1), 1–8.

Pepe, E., Bajardi, P., Gauvin, L., Privitera, F., Lake, B., Cattuto, C., & Tizzoni, M. (2020). Covid-19 outbreak response: A first assessment of mobility changes in Italy following national lockdown. *MedRxiv.*

Reynolds, D. A. (2009). Gaussian mixture models. *Encyclopaedia of biometrics, 741*, 659–663.

Scherrer, L., Tomko, M., Ranacher, P., & Weibel, R. (2018). Travellers or locals? Identifying meaningful sub-populations from human movement data in the absence of ground truth. *EPJ Data Science, 7*(1), 19.

Shin, H., Chon, Y., & Cha, H. (2012). Unsupervised construction of an indoor floor plan using a smartphone. *IEEE Transactions on Systems, Man, and Cybernetics, Part C (Applications and Reviews), 42*(6), 889–898.

Siła-Nowicka, K., Vandrol, J., Oshan, T., Long, J. A., Demšar, U., & Fotheringham, A. S. (2016). Analysis of human mobility patterns from GPS trajectories and contextual information. *International Journal of Geographical Information Science, 30*(5), 881–906.

Solmaz, G., & Turgut, D. (2019). A survey of human mobility models. *IEEE Access, 7,* 125711–125731.

Solomon, A., Bar, A., Yanai, C., Shapira, B., & Rokach, L. (2018). Predict demographic information using word2vec on spatial trajectories. In *Proceedings of the 26th Conference on User Modeling, Adaptation, and Personalization* (pp. 331–339). New York: ACM.

Ta, N., Zhao, Y., & Chai, Y. (2016). Built environment, peak hours and route choice efficiency: An investigation of commuting efficiency using GPS data. *Journal of Transport Geography, 57,* 161–170.

Tang, J., Liu, F., Wang, Y., & Wang, H. (2015). Uncovering urban human mobility from large scale taxi GPS data. *Physica A: Statistical Mechanics and its Applications, 438,* 140–153.

Tian, X., Wu, X., Li, H., & Wang, X. (2019). Rf fingerprints prediction for cellular network positioning: A subspace identification approach. *IEEE Transactions on Mobile Computing, 19*(2), 450–465.

Urban Redevelopment Authority (URA) (2019). *Master plan. Boundary data.* Retrieved from URA Master Plan 2019, https://data.gov.sg/dataset/master-plan-2019-subzone-boundary-no-sea.

Van der Spek, S., Van Schaick, J., De Bois, P., & De Haan, R. (2009). Sensing human activity: GPS tracking. *Sensors, 9*(4), 3033–3055.

Wang, Y., Qin, K., Chen, Y., & Zhao, P. (2018). Detecting anomalous trajectories and behaviour patterns using hierarchical clustering from taxi GPS data. *ISPRS International Journal of Geo-Information, 7*(1), 25.

Wu, C., Yang, Z., & Liu, Y. (2014). Smartphones based crowdsourcing for indoor localization. *IEEE Transactions on Mobile Computing, 14*(2), 444–457.

Wu, L., Yang, L., Huang, Z., Wang, Y., Chai, Y., Peng, X., & Liu, Y. (2019). Inferring demographics from human trajectories and geographical context. *Computers, Environment and Urban Systems, 77,* 101368.

Xu, Y., Belyi, A., Bojic, I., & Ratti, C. (2018). Human mobility and socioeconomic status: Analysis of Singapore and Boston. *Computers, Environment and Urban Systems, 72,* 51–67.

Zheng, Y., Li, Q., Chen, Y., Xie, X., & Ma, W.-Y. (2008). Understanding mobility based on GPS data. In *Proceedings of the 10th International Conference on Ubiquitous Computing* (pp. 312–321). New York: ACM.

Zheng, Y., Zhang, L., Xie, X., & Ma, W.-Y. (2009). Mining interesting locations and travel sequences from GPS trajectories. In *Proceedings of the 18th International Conference on World Wide Web, ser. WWW '09* (pp. 791–800). New York: Association for Computing Machinery.

Zhou, P., Zheng, Y., Li, Z., Li, M., & Shen, G. (2012). Iodetector: A generic service for indoor outdoor detection. In *Proceedings of the 10th ACM Conference on Embedded Network Sensor Systems* (pp. 113–126). New York: ACM.

Zhou, Y., Lau, B. P. L., Koh, Z., Yuen, C., & Ng, B. K. K. (2020). Understanding crowd behaviours in a social event by passive wifi sensing and data mining. *IEEE Internet of Things Journal, 7*(5), 4442–4454.

Zhu, J. Y., Zheng, A. X., Xu, J., & Li, V. O. (2014). Spatio-temporal (ST) similarity model for constructing Wi-Fi-based RSSI fingerprinting map for *indoor localization. In Indoor Positioning and Indoor Navigation (IPIN), 2014 International Conference on* (pp. 678–684). IEEE.

Zhu, L., Gonder, J., & Lin, L. (2017). Prediction of individual social-demographic role based on travel behaviour variability using long-term GPS data. *Journal of Advanced Transportation, 2017,* 1–13.

# 5

# LARGE-SCALE HUMAN MOBILITY TRACKING IN RESIDENTIAL ESTATES

*Zann Koh, Hasala Marakkalage, Billy Pik Lik Lau, Yuren Zhou, Ran Liu and Chau Yuen*

## Introduction

Chapter 4 described some methods of using GPS data to extract insights into the mobility of individual users. However, using GPS tracking may be less feasible when trying to analyse the mobility of a crowd or large groups of users. In such cases, passively sensed signals detected using installed sensors can be used instead to track the mobility of a large group of users within a fixed study area. Examples of passively sensed smartphone signals that can be used in mobility tracking are Bluetooth signals and Wi-Fi signals. These signals can be detected while being emitted from users' smartphones without the need for users to actively participate.

The case study in this chapter presents selected methods for the analysis of residents' travel behaviours within a single housing estate as well as the comparison of residents' travel behaviours between two housing estates. The data from this study was gathered via several Wi-Fi sensors deployed around the estates of study. For residents' behaviours within a single estate, different hidden patterns were explored such as how the residents' activeness change over time of day, how day type (DT) and rain affect the residents' activeness, as well as how residents move around between different parts of the estate. The comparison of residents' travel behaviours between two housing estates was performed through a proposed feature extraction and clustering technique.

## Case Study Background

Residential estates play an important role within any social community as homes are where people spend most of their time. Therefore, for urban planning, it is essential to consider the mobility of residents and usage of facilities within and

DOI: 10.4324/9781003437659-8

around existing residential estates to find points of potential improvement for the planning of future estates.

This case study aims to track and analyse the differences in mobility data within two residential estates of different neighbourhood design typologies. The tracking of such a large group of users within these fixed study areas requires large-scale passive data collection methods that can capture the patterns of many passers-by simultaneously. Two possible methods are passive Bluetooth sensing and passive Wi-Fi sensing. Out of these two, passive Wi-Fi sensing was chosen for this study as Bluetooth detection records have a much smaller detection rate, which can be as low as 3% of Wi-Fi probe request records (Schauer, Werner, & Marcus, 2014).

Passive Wi-Fi sensing involves sensing the probe requests sent out by nearby smart devices that search for an available Wi-Fi connection. These probe requests contain some information about the devices that send them, including the Media Access Control (MAC) numbers, which can be used to group trajectories made by a single device and reduce occurrences of double counting. Passive Wi-Fi sensing has been used in past mobility studies with various scenarios including large-scale events (Zhou, Lau, Koh, Yuen, & Ng, 2020; Alessandrini, Gioia, Sermi, Sofos, Tarchi, & Vespe, 2017), in large public facilities (Prentow, Ruiz-Ruiz, Blunck, Stisen, & Kjærgaard, 2015; Zakaria, Trivedi, Cecchet, Chee, Shenoy, & Balan, 2020), as well as within residential estates (Koh, Zhou, Lau, Yuen, Tuncer, & Chong, 2020), (Zhou, Koh, Ng, Liu, Foo, Yuen, & Chong, 2021).

One important use of passive Wi-Fi sensing is to estimate device counts across time as a measure of human activity near defined sensor locations. According to Ribeiro et al. (Ribeiro, Nunes, Nisi, & Schöning, 2020), it is possible to use the counts of detected devices after fine-tuning to estimate the actual number of people. In this manner, passive Wi-Fi sensing has been used to estimate occupancy in different locations such as retail stores (Depatla & Mostofi, 2019), smart buildings (Ciftler, Dikmese, Gᴚuvenc, Akkaya, & Kadri, 2017), and on public transport (Pu, Zhu, Li, Cui, Guo, & Wang, 2020).

## System Overview

To investigate such temporal variations of human activity within residential estates, this study consists of two portions. The first portion addresses the investigation of human activity patterns in a single estate, while the second part proposes a method which can be used to compare the activity levels of two estates. The data flow within the framework of this study is illustrated in Figure 5.1.

For the single-estate study, an initial visualisation is performed on the counts of unique devices separated by MAC addresses over different times of day. The machine learning technique of linear regression is then adopted to investigate probable factors that affect these activity levels, which include DT and rain. Next, two transition probability matrices are computed from extracted trajectories, which represent where an individual is likely to go towards from a given location and to

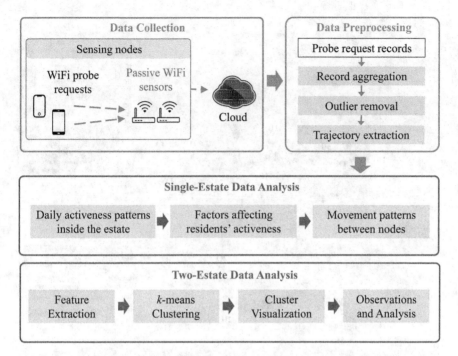

**FIGURE 5.1**    Overall framework for understanding residents' behaviours through passive Wi-Fi sensing and data mining.

*Source*: Authors.

come from to reach a given location. The two matrices were then subjected to hierarchical agglomerative clustering (HAC) (Müllner, 2011) to group locations based on their interconnectivity.

For the multi-estate study, data from two estates was used. A feature extraction method called smoothing, peak detection, and clustering (SPDC) is proposed, where intuitive features of each day's activity patterns at each location are extracted and clustered to discover groups of similar locations based on the data. The identified clusters were then analysed based on their composition of location types and DTs. These can help to identify possible links or differences between locations with similar functions in two different residential estates.

## Data Collection

Data for this study was collected from two housing estates in Singapore, namely, Topaz and Cascadia, which are adjacent estates in the town of Punggol, referred to from here onwards as Estate 1 and Estate 2. The two estates were selected because they are representative of distinctive neighbourhood design typologies: Estate 1 is a conventional estate with a standalone multi-storey carpark building and with public

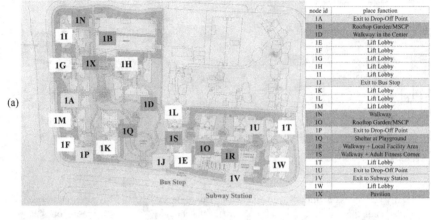

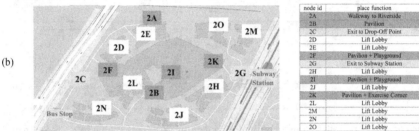

**FIGURE 5.2** Maps of sensor locations at (a) Estate 1: Topaz and (b) Estate 2: Cascadia. The nodes are colour-coded by type: white is for lift lobbies, yellow for entrances/exits, and green for public spaces such as playgrounds, FCs, pavilions, and walkways.

*Source*: Authors.

facilities all on the ground level, whereas Estate 2 is a newer design with housing blocks sitting on top of a two-storey multi-storey car park (MSCP) which forms an environmental deck on the roof of the MSCP. Within each estate, the locations of interest such as lift lobbies, exits, and public facilities such as playgrounds and walkways were identified and sensor boxes, referred to as nodes, were deployed at each location of interest. The diagrams of the deployed sensor locations and location types are shown in Figure 5.2. The nodes at Estate 1 were labelled from 1A to 1X, while the nodes at Estate 2 were labelled from 2A to 2O.

From Estate 1, around 34.7 million raw records were collected over a period of 55 days, while in Estate 2, about 9.84 million raw records were collected over a period of 83 days. The difference between the rate of records collected between the two estates may be due to Estate 1 having more residential buildings over a larger land area as well as more ground-level amenities.

The data collected by the sensors when detecting Wi-Fi probe requests from nearby devices include the date and time of detection, the MAC address of the detected device, as well as the identifier (ID) of the detecting sensor. No other identifiable information is recorded for the preservation of privacy.

## Data Processing

The raw data must be processed before further analysis can be done. First, the counts of each unique MAC address at each node were aggregated in intervals of 15 minutes, such that it was neither too long, which causes excessive information loss, nor too short, which may lead to noise overwhelming the pattern. Other literature regarding daily temporal patterns ((Zhou, Koh, Ng, Liu, Foo, Yuen, & Chong, 2021), (Graells-Garrido, Ferres, Caro, & Bravo, 2017), (Kalogianni, Sileryte, Lam, Zhou, Van der Ham, Van der Spek, & Verbree, 2015)) have used a range of different aggregation intervals, which includes 15 minutes as well. When the data is aggregated in this manner, each combination of node location and date of data collection can be represented by a vector with 96 values representing the number of detected devices within each 15-minute interval of the day. The values of each date are tracked from 3 am on that date to 3 am the next calendar day to minimise the spillover effects of activity near midnight. Each such vector, referred to as a date-node vector, can be plotted as a line to show the evolution of device count at a specific node over the course of a specific date.

Raw data also needs to be cleaned as it may contain noise. For this study, there arose two cases where data needed to be replaced. The first case was missing data. A check was performed to determine if the date-node vector had more than 50% of its data missing. If it did, that date-node vector was removed from the dataset for that specific day. If less than 50% of the data was missing, the missing values were then imputed using the median value from the remaining date-node vectors from the same node at the same time slot of the missing values. For example, if there was data missing at 1 pm for a specific vector, it would be replaced with the median values calculated from the 1 pm timeslot of the rest of the vectors from the same node on other days.

The second case in which data needed to be replaced occurred when data was unusually high or low for a certain time slot at a certain node, which means they were outliers. The thresholds for such outliers were calculated by using the equations below:

$$upper = 1.5 \times (p_{75} - p_{25}) + p_{75}$$
$$lower = p_{25} - 1.5 \times (p_{75} - p_{25})$$
(1)

where $p_{75}$ and $p_{25}$ are the 75th and 25th percentile of the values of a specific timeslot over all the valid dates at a certain node. The thresholds were calculated separately for different timeslots and different nodes. If the outlier was higher than the upper threshold value, it was replaced with the upper threshold value, whereas if the outlier was lower than the lower threshold, it was replaced with the lower threshold value.

For investigating the likelihood of travel between locations, the trajectory data of each MAC address was extracted for each day. This is done by firstly sorting all probe request records corresponding to the same MAC address in a single day by time of appearance. Next, records which are chronologically adjacent with the same node ID and an interval of less than five minutes were combined as a single-node visit. The possibility of temporal conflicts between two sensing nodes was present as some nodes were deployed in outdoor environments, where the Wi-Fi range is larger and hence signals from two nodes may overlap. The rules proposed in a previously published paper (Zhou, Lau, Koh, Yuen, & Ng, 2020) were applied to solve such temporal conflicts. Finally, extremely short trajectories and those from estate workers or fixed devices were filtered and removed based on their duration and count of visited nodes within each estate.

## Single-Estate Data Analysis

The study of a single estate aims to extract insights for three important aspects that could be of interest to the estate manager and designer: (1) daily activeness patterns inside the estate, (2) factors affecting the residents' activeness, (3) movement patterns between nodes. For this study, the data from Estate 2 was used as it covered a longer duration for each node.

### Daily Activeness Visualisations

The date-node vectors obtained from data processing were aggregated into four different DTs as shown in Table 5.1. The device count in each 15-minute interval is averaged within each of these DTs to get a typical daily device count curve for the corresponding type. These four different DTs are proposed based on prior research (Koh, Zhou, Lau, Yuen, Tuncer, & Chong, 2020) that discovered that these DTs may show different types of activity.

For an overview of the entire estate's activity on each of these four different DTs, the typical daily device count curve is plotted in Figure 5.3. Obvious differences can be observed between the different DTs. The working days (DT1 and DT2) show a morning peak of activeness, which likely represents residents going

**TABLE 5.1** Definition and count of days of each DT

| Day Types | Definition | Count of Days |
|-----------|------------|---------------|
| DT1 | Workdays excluding Fridays and PH eves | 45 |
| DT2 | Working Fridays and PH eves | 12 |
| DT3 | Saturdays and PHs with the next day off | 14 |
| DT4 | Sundays and PHs with the next day working | 12 |

*Source*: Authors.

*PH*: Public Holidays.

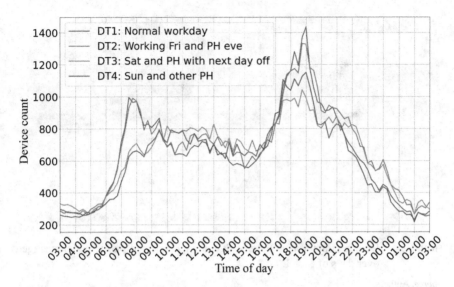

**FIGURE 5.3** Device count versus time of day in the entire estate, averaged over days of each DT.

*Source*: Authors.

out for work. During the daytime, between 10 am and 4 pm, the overall detected activeness is less on workdays as compared to rest days (DT3 and DT4). The estate is most active for all four of the DTs in the evenings between 5 pm and 8 pm, while workdays witness higher activity than rest days, which may be due to students and professionals returning home. Comparing DT1 and DT2, one can find that Fridays and Public Holiday (PH) eves have higher activeness after 9 pm than other workdays. This may indicate that residents are out and about until later at night since the next day is a rest day. A similar comparison may also be made between DT3 and DT4, which suggests that Saturdays and PHs with the next day off have higher activeness at night than the other rest days.

Beyond the overall activity in the estate, the temporal activity levels in each estate can be investigated individually as well, which can inform estate managers and designers about how much each public facility is utilised. The daily device count curves of six selected public nodes are thus plotted in Figure 5.4 for visualisation.

There are two exit nodes within the six public nodes, namely, node 2C, which leads to the bus stop, and node 2G, which leads to a subway station. A comparison of their daily curves from Figure 5.4 shows that node 2G has a sharp morning peak between 6 am to 8 am on working days (DT1 and DT2) that does not appear for rest days (DT3 and DT4), whereas no such distinction appears in the graphs of node 2C. This could indicate that the subway station is a more popular choice for commuting to work in the mornings for the residents, and thus the urban designers may have to make accommodations for the morning rush or facilitate the movement between residential blocks which are further away from this gate.

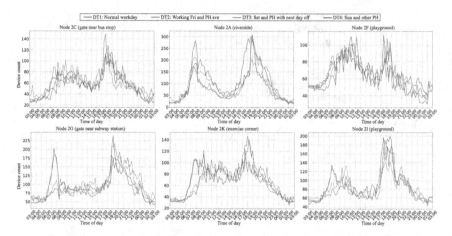

**FIGURE 5.4**   Device count versus time of day at each selected public node, averaged over days of each DT.

*Source*: Authors.

The next two nodes are node 2A (riverside path) and node 2K (exercise corner). On a functional level, both nodes may be used for workout purposes. However, their activity level patterns are different. The highest activity levels sensed by both nodes include a morning peak between 6 am to 9 am and an afternoon peak between 4 pm and 7 pm. However, these peaks are the sharpest and highest on Sundays and PHs (DT4) for node 2A, while for node 2K they are not as sharp, and the highest points occur on working days (DT1 and DT2). The morning and evening peaks on working days for node 2K may be caused in part by its location next to the exit node 2G to the subway; thus, it may detect the rush of commuters from other buildings passing by in the morning and evening.

The last two nodes are located near playgrounds inside the estate, which are nodes 2F and 2I. For node 2F, its usage rate has minimal difference across the four different DTs, and higher levels occur between 7 am to 1 pm and 5 pm to 7 pm. In contrast, the usage rate at node 2I has different active periods in the morning for working days (6 am–8 am) and rest days (8 am–10 am). The activity levels for node 2I are mostly much higher in the evening compared to the morning, while the high activity levels for node 2F are about the same level in the morning and the evening. Node 2I also shows a moderate increase in activity after 8 pm, which may indicate casual after-dinner activities.

## Factors Affecting Residents' Activeness

After visualising the daily activeness patterns, the observation may be made that the DT can affect the activity levels of the entire estate and some public nodes significantly. The different DTs (DT1 to DT4) may be further generalised

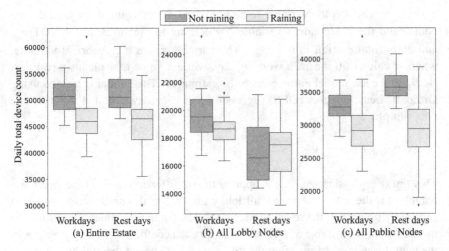

**FIGURE 5.5** Box plots of daily total device count under different conditions of DT and rainfall for (a) the entire estate, (b) all lift lobby nodes, and (c) all the public nodes. A day is considered as raining when its daily total rainfall (mm) is more than 0.

*Source*: Authors.

into working days (DT1 and DT2) and rest days (DT3 and DT4) as it was observed that the main DT-based differences in activity levels were due to this distinction. Another factor which may significantly affect activity levels includes rain, as many of the nodes are located outdoors. For the rainfall data, the daily total rainfall (mm) of the studied area is obtained from the website of Meteorological Service Singapore (MSS) (MSS Historical daily weather records, 2021–03–17).

To get an initial sense of how much each of these factors affects the activeness levels, box plots of daily total device count under different conditions of DT and rainfall were plotted in Figure 5.5. The box plots represent the device count values for three different groups of nodes – the entire estate, all lift lobby nodes, and all public nodes. Figure 5.5a shows a clear reduction of daily activity level in the presence of rain over the entire estate, while the DT of workday or rest day seems to have minimal impact on the activity levels. For lobby nodes in Figure 5.5b, the effect of DT on the activity level seems to be clearer than the impact of rain, which is reasonable as the lobby nodes are all enclosed under shelter in the housing blocks. In contrast, Figure 5.5c shows that the activity levels at public nodes, which are deployed outdoors, are more affected by the presence of rain, which causes a lower activity level.

For a more specific quantitative description of the impact of DT and rainfall on residents' activity levels, linear regression analysis (Kleinbaum, Kupper, Nizam, & Rosenberg, 2013) is conducted as shown in Equation 2. The dependent variable $y$, which is the variable to be predicted, is set as the daily total device count, and the independent variables, which are the features, are the DT $x_{off}$ and daily total rainfall in mm $x_{rain}$. The variable $x_{off}$ is a categorical variable, where a value of 0 means a working day, while a value of 1 means a rest day. $\beta_1$, $\beta_2$, and $c$ are model parameters to be estimated. This model is trained using Ordinary Least Squares (OLS) regression and each day's data is input as one data sample.

$$y = \beta_1 x_{off} + \beta_2 x_{rain} + c \tag{2}$$

This linear regression analysis is applied to ten different cases. These cases are, namely, for the entire estate, for all lobby nodes, for all public nodes, and then separately for each of the seven public nodes. The computed values after fitting the model are shown in Table 5.2. The $p$-value for each coefficient $\beta_1$ and $\beta_2$ represents the confidence level of rejecting the null hypothesis, where the null hypothesis is that the value of the coefficient is zero, meaning that the corresponding independent variable has no effect on $y$. A small $p$-value of less than the commonly used threshold of 0.05 shows that there is a high confidence that the null hypothesis is rejected and thus the corresponding independent variable has a statistically significant impact on the dependent variable. The significant $p$-values and corresponding coefficient values are marked in red.

The observations drawn from Figure 5.5 are quantitatively supported by the given parameter values in the first three cases of the entire estate, all lobby nodes, and all

**TABLE 5.2** Calculated $p$-values and real values of $\beta_1$ and $\beta_2$

|  | $x_{off}$ | | $x_{pain}$ | |
| --- | --- | --- | --- | --- |
|  | p-value | $\beta_1$ | p-value | $\beta_2$ |
| Entire estate | 0.493 | 0.138 | 0.000 | −0.663 |
| All lobby nodes | 0.000 | −0.892 | 0.342 | −0.103 |
| All public nodes | 0.008 | 0.512 | 0.000 | −0.738 |
| Node 2A (riverside) | 0.000 | 0.902 | 0.000 | −0.684 |
| Node 2B (pavilion) | 0.715 | 0.090 | 0.004 | −0.342 |
| Node 2C (exit near bus stop) | 0.168 | 0.326 | 0.000 | −0.475 |
| Node 2F (playground) | 0.593 | −0.128 | 0.001 | −0.386 |
| Node 2G (exit near subway station) | 0.313 | −0.259 | 0.263 | −0.134 |
| Node 2I (playground) | 0.217 | 0.275 | 0.000 | −0.563 |
| Node 2K (exercise corner) | 0.008 | −0.616 | 0.005 | 0.306 |

*Source*: Authors.

public nodes. For the individual public nodes, nodes 2A and 2K are the only ones affected by both the rain and DT factors. The rain has a negative effect on the activity level for both nodes, but the rest days increase the activity level at node 2A while decreasing the activity level at node 2K. The effect of DT on these two nodes here agrees with the observations from Figure 5.4. The other public nodes 2B (pavilion), 2F (playground), and 2I (playground) are affected only by the presence of rain and not DT, which may form a motivation for the estate managers to provide indoor facilities to residents in the event of rain. Lastly, the two exit nodes (2C and 2G) are not significantly affected by the DT. However, the activity levels at node 2C are slightly affected on rainy days.

## Movement Patterns between Nodes

Other than investigating the patterns in activity levels of individual nodes, the movement of residents between different nodes is another important consideration for facility planning. For this investigation, trajectories of individual devices are extracted based on the MAC address as described in the Data Processing subsection. A transition count matrix $N$ is generated by counting the number of times the residents travel between each pair of nodes in their trajectory sequences. Each row of $N$ represents the origin node from 2A to 2O and the column represents the destination node from 2A to 2O as well. The value of each element of $N$, denoted as $N(i, j)$, would thus represent the total number of times that all the residents were detected going from node $i$ to node $j$ during the collection period.

Next, the transition count matrix $N$ is normalised by its rows to obtain the transition probability matrix $P_{to}$. In a similar manner, $N$ is also normalised by its columns to obtain the transition matrix $P_{from}$. The descriptions of these matrices can be found in Equation 3. The value $P_{to}(i, j)$ represents the probability that a person located at node $i$ would go to node $j$, while $P_{from}(i, j)$ represents the probability that a person located at node $j$ has come from node $i$. When $i = j$, the values of $P_{to}(i, j)$ and $P_{from}(i, j)$ are set as zero to focus on trips between different nodes.

$$P_{to}(i, j) = \frac{N(i, j)}{\sum_{k \in [1,10], k \neq i} N(i, k)}$$

$$P_{from}(i, j) = \frac{N(i, j)}{\sum_{k \in [1,10], k \neq i} N(k, j)} \tag{3}$$

The interconnections of the nodes based on the transition probability matrices can be automatically described using HAC (Müllner, 2011), where each matrix is input as a feature table separately. The matrix $P_{from}$ is transposed before inputting to ensure that its rows are normalised as each row represents the features of a node. HAC will then organise the given data points into a hierarchical tree-like structure called a dendrogram.

The levels observed on the obtained dendrogram indicate the level of interconnection between data points. The points are iteratively agglomerated from the bottom, where they are individual nodes, to the top, where the groups increase in size and eventually the whole set of points is in one group. Clusters can be extracted from these dendrograms by selecting a level of agglomeration to stop at and observing which nodes are agglomerated together below that level.

To measure the feature similarity between each pair of nodes, Euclidean distance was selected. As the feature similarity represents the interconnection degree between each pair of nodes, a smaller Euclidean distance indicates a stronger interconnection level. When nodes are agglomerated into clusters, each cluster's interconnection value to another cluster is represented by the average of the interconnection degree (similarity) between all pairs of nodes in the two clusters. The average value is computed using Equation 4.

$$L(C_1, C_2) = \frac{1}{|C_1| \cdot |C_2|} \sum_{a \in C_1} \sum_{b \in C_2} \sqrt{\sum_l (a_l - b_l)^2} \tag{4}$$

In Equation 4, $L(C_1, C_2)$ stands for the linkage criterion between two clusters of nodes $C_1$ and $C_2$. $|\cdot|$ represents the cardinality of each cluster. The variables $a$ and $b$ represent the nodes belonging to $C_1$ and $C_2$, respectively, and $l$ stands for the index of each feature of each node (i.e., the $l$th value of each node in $P_{to}$ and $P_{from}^T$).

The dendrograms obtained from HAC are plotted in Figure 5.6. The HAC process orders the leaf nodes of the tree such that the distance between successive leaves is minimal (Bar-Joseph, Gifford, & Jaakkola, 2001). From the figure, the two sets of input features $P_{to}$ and $P_{from}^T$ result in an identical grouping of nodes, despite some minor differences of the node order.

Reading from the bottom of the dendrograms upwards, some pairs of nearby lobby nodes (2L and 2B, 2J and 2H, 2D and 2E, 2O and 2M) are clustered together first. Some public nodes such as 2C and 2F, and 2I and 2K, are also paired. Nodes 2O and 2M are grouped with node 2G on the next level, which may be because the path leading from 2O and 2M to other nodes passes by 2G first. Moving upwards along the dendrogram, the entire estate can be roughly divided into four sub-areas containing the respective nodes: (1) 2N, 2C, 2F, (2) 2L, 2B, 2I, 2K, 2J, 2H, (3) 2D, 2E, 2A, and (4) 2O, 2M, 2G.

For a higher-level examination of the transitions throughout the estate, the 15 nodes within the estate are aggregated into ten zones as shown in Figure 5.7. These zones are decided based on the lower-level groupings of the nodes from Figure 5.7. Nodes 2C and 2G are not grouped with any other nodes as they are located at the gates of the estate.

Next, the transition probability matrices $P_{to}^{zone}$ and $P_{from}^{zone}$ are computed using Equation 4, where the individual node counts are replaced with zone counts. These matrices are visualised as heatmaps in Figure 5.8. Each row in Figure 5.8a shows the zone where a person is more likely to go to next from their current zone. For example, Zone

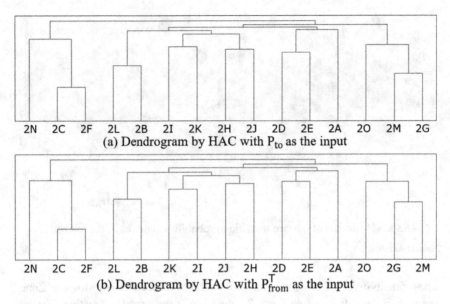

**FIGURE 5.6** Dendrograms obtained from HAC using the two transition probability matrices as input.

*Source*: Authors.

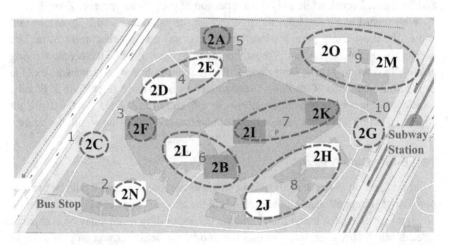

**FIGURE 5.7** Ten zones (red circles) obtained by grouping some of the sensing nodes based on the HAC dendrograms.

*Source*: Authors.

1 (2C) has a high likelihood of going to Zone 3 (2F) next, and vice versa. This could be caused by the proximity of node 2F to 2C, where node 2C is near an exit of the estate and thus the people who enter or exit near node 2C may pass by node 2F frequently.

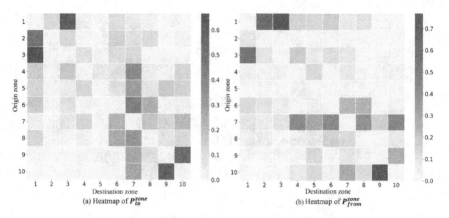

**FIGURE 5.8**   Heatmaps of the two transition probability matrices of the ten zones.

*Source*: Authors.

The residents moving from Zone 2 have a higher probability of moving towards Zone 1 rather than towards Zones 6, 7, and 8, which are in the opposite direction. The residents leaving from Zones 4, 6, and 8 mainly tend to go to Zone 7 as the next step. This can be explained by the central location of Zone 7 with respect to the estate boundaries, as well as its proximity to the gate node 2G. The dominant next steps of Zones 9 and 10 are reciprocal, while a significant portion of people also travel to Zone 7.

With regard to Figure 5.8b, the sum of each of its columns adds up to one and shows the most likely previous location of a person given their current zone. The people arriving at Zones 2 and 3 mostly come from Zone 1, while the people arriving at Zone 1 commonly come from Zone 3. This may indicate that residents who exit the estate from gate node 2C at Zone 1 tend to pass through node 2F at Zone 3. Finally, most of the zones with lobby nodes (Zones 4, 5, 6, and 8) have Zone 7 as the dominant source, meaning that Zone 7 could be a popular transitory location within the estate.

## Two-Estate Data Analysis

After performing in-depth analysis on a single estate, the next step would be to compare two different estates. All the methods in the above section can be performed on any number of estates individually and the corresponding results may be compared to each other. However, some analysis methods can be performed on two or more estates at the same time. One such method is described in the next case study, which uses the passively sensed Wi-Fi data from two estates collected and processed as described in the earlier section "System Overview". Previously, the date-node vectors obtained were used for visualisation as well as linear regression analysis. In this two-estate data analysis, these vectors act as the starting points from which features are extracted and used for clustering and insight extraction. The feature extraction and clustering methodology is a three-step process involving SPDC.

## Smoothing

As with most real-world data, the collected data has noise which may affect the results of any machine learning technique. This study focuses on extracting the main peak features over each single day; therefore smoothing is required. An example of a plot with noise is shown in Figure 5.9a. If the algorithm simply detected the highest three peaks from this noisy data, the result would be r1, r2, and r3, which may not be a good indicator of the overall day's patterns as the slight peak in the middle, which may be something important, would be ignored. After smoothing, the three main peaks are detected as s1, s2, and s3, which may describe the day's patterns more fully.

For smoothing, LOcally WEighted Scatterplot Smoothing (LOWESS) (Cleveland, 1979) was utilised. The process of LOWESS uses a fraction, $\alpha$, of the nearest data points surrounding each data point of smoothing to fit a low-level polynomial to each data point. For this study, a value of $\alpha = 0.2$ was found to be most suitable from manual testing.

## Peak Detection

As most of the days in the data used for this case study contained three main peaks or less, the intuitive features from the first three peaks were extracted as features for clustering. The peaks are first detected using the following definition: For an ordered sequence $\left[ x_1, x_2, \ldots, x_i, \ldots, x_n \right]$ where $n$ is the total number of values in the sequence, a peak is detected at location $i$ where both of the following apply: $x_{i-1} < x_i$ and $x_{i+1} < x_i$.

Three features were extracted from each peak: peak height $h$, peak width at half height $w$, and peak locations on the $x$-axis $l$, which ranges from 1 to 96 for each 15-minute interval. These features are illustrated in Figure 5.9(b). If there are less than three peaks occurring in a single day, the remaining peak features are set to zero. The extracted feature vector would be in the form of $\left\{ h_1, w_1, l_1, h_2, w_2, l_2, h_3, w_3, l_3 \right\}$ where each subscript represents the ranking of the peak height, from one being the

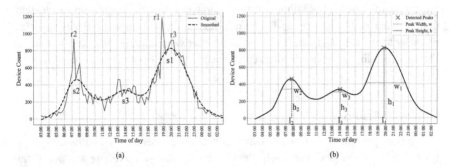

(a) (b)

**FIGURE 5.9** (a) Comparison of the raw data before and after smoothing. (b) Extracted features from the smoothed data. The peak location, l, is given by the position on the $x$-axis corresponding to the detected peaks.

*Source*: Authors.

tallest peak to three being the shortest. In theory, this would help the clustering algorithm to group days containing peaks with similar shapes and locations.

Before clustering, min-max normalisation was performed on all the extracted feature vectors to ensure that all the features are of the same scale, to be able to compare them between various locations and time frames. The min-max normalisation uses the following formula:

$$X' = \frac{(X - min(X))}{(max(X) - min(X))} \tag{5}$$

## Clustering

As the data generated via passive Wi-Fi sensing comes in large volumes, an effective method to extract insights from such large volumes of data is required. This can be achieved by using machine learning techniques such as clustering. A plethora of clustering algorithms is available in the literature, each with various requirements and parameters. Some algorithms are better suited for use with low-dimensional datasets, like Mean Shift Clustering (Comaniciu & Meer, 2002) and Balanced Iterative Reducing and Clustering using Hierarchies (BIRCH) (Zhang, Ramakrishnan, & Livny). Others require parameters which may be difficult to fine-tune, such as Density-Based Spatial Clustering of Applications with Noise (DBSCAN) (Ester, Kriegel, Sander, Xu, et al., 1996), which requires both a noise tolerance parameter and a threshold distance parameter; $k$-means clustering (Lloyd, 1982) was finally selected as it is easily implemented and scalable to large amounts of data, which is required for the data used in this study.

For $k$-means clustering, an input parameter $k$ is required. To find an optimal value for this parameter, two cluster validity indexes were used, the silhouette index (Rousseeuw, 1987) and the Davies-Bouldin index (Davies & Bouldin, 1979). The silhouette index uses the mean distance within each cluster as well as the mean distance to its nearest cluster over all the data points in the cluster to assess the cluster validity. A higher silhouette score indicates a better separation of clusters. On the other hand, the Davies-Bouldin index uses the average similarity measure of each cluster with its nearest cluster, so a better separation is indicated by a lower Davies-Bouldin score. Both indexes showed that the optimal value of $k$ for this case is 17.

## Clustering Results

Using the optimum value of 17, $k$-means clustering was performed using the extracted feature vectors. For each cluster, the normalised average temporal pattern and the extents of the 25th and 75th percentile of the cluster are plotted in Figure 5.10. Visually, the clusters can be grouped into four larger groups based on their defining characteristics: Group 1 has two peaks with heights that are comparable, Group 2 has a higher morning peak than the evening peak, Group 3 has a

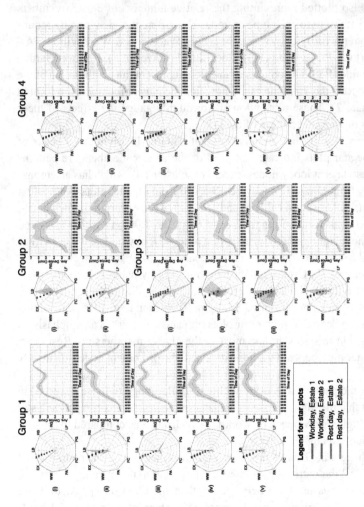

**FIGURE 5.10**  Visualisations of the 17 clusters of daily device count patterns derived from *k*-means clustering. The 17 clusters are grouped into four groups based on the relative peak heights and peak numbers. The Roman numerals (i) to (vi) are used as an index to refer to a specific cluster within a group. The node type labels are as follows – Lobby (LB), Exit (EX), Walkway (WW), Pavilion (PA), Fitness Corner (FC), Playground (PG), Local Facility Area (LF), Rooftop Garden (RG), and Riverside (RS).

*Source:* Authors.

higher evening peak than a morning peak, and Group 4 has the clusters with three peaks. Within each group, the Roman numerals (i) to (vi) are used to denote specific clusters detected by the algorithm.

In the previous single-estate data analysis, the type of day, namely, Workday or Rest day, was identified to be a potential factor that affects the daily temporal activity patterns. The function of each node may also be a similar factor. Hence, the star plots were also plotted representing the relative numbers of cluster members in each cluster pertaining to each type of node. The different node types are (counterclockwise from the top) Lobby (LB), Exit (EX), Walkway (WW), Pavilion (PA), Fitness Corner (FC), Playground (PG), Local Facility Area (LF), Rooftop Garden (RG), and Riverside (RS). The colours of the plots represent the combination of the DT (Workday or Rest day) and the estate in which the cluster members are located (Estate 1 or Estate 2). These plots allow us to roughly gauge the effects of these different factors on the final pattern of the DT.

The clusters in Group 1 are mainly those of patterns involving two peaks of similar size. The star plots indicate that most of the cluster members are patterns occurring on Rest days at lobby nodes, except for cluster 1(i), which has an emphasis on exit nodes from Estate 1. It may be observed from the line plots that the first peak location varies between 11:00 am and 1:00 pm, which is around lunch hours, while the second peak location ranges from 5:00 pm to 7:00 pm, which is around dinner hours. The insight gathered here is that at these lobby nodes, the activity is higher on Rest days around lunch and dinner hours.

For the two clusters in Group 2, their temporal activity patterns both start with an increase in activity until a peak at around 8:30 am, followed by a decrease in activity at different rates until around 3:00 pm. The second, smaller peak occurs around 6:30 pm. The corresponding star plots indicate that there is an emphasis on playground nodes in this cluster, and most of the cluster members are Rest days. More of the cluster members are from Estate 2, and the prominent nodes are mostly public spaces.

For Group 3, the peak locations are within a similar range to those observed in Group 2, except that the higher peak occurs later in the day. The corresponding star plots indicate a higher prevalence of Workdays in Group 3 as compared to those in Group 2, especially in cluster 3(iii).

Lastly, the clusters in Group 4 are those with three peaks over the course of the day. In Group 4, all the patterns have their highest peak in the evening. Their first peak in the morning varies from being lower than the afternoon peak, to being much higher than the afternoon peak. The associated star plots show that these patterns mostly occur at lobby and/or exit nodes, and mostly on Workdays.

Overall, Groups 2 and 3 involve a higher proportion of nodes at public spaces as compared to Groups 1 and 4, which focus more on lobby and exit nodes. From Groups 2 and 3, it may be generalised that public spaces are more likely to have two peaks of activity throughout the day, one in the morning and one in the evening, while Groups 1 and 4 indicate that lobby and exit nodes mostly have two peaks on Rest days and three peaks on Workdays. It may also be observed that Group 1 has more of a focus on Estate 2 while Group 4 has more of a focus on Estate 1.

Focusing on the star plots also allows for the observation of the differences in activities of different DTs for the same node for specific nodes such as 2A, which is the only node labelled as RS: Riverside. Most of the Rest days at this node are in cluster 2(i) with a higher morning peak, in contrast to its Workdays, which are in cluster 3(ii) with a higher evening peak. This may indicate that users visit the riverside area more frequently after work on Workdays and earlier in the morning on Rest days to avoid the harsh afternoon sun.

Next, for ease of visualisation of any potential spatial factors, the maps in Figure 5.11 show the common nodes that are linked to more than ten members of each cluster. Some of the clusters may be too small or spread out across many nodes, and thus may not have circles marked out.

The plots in Figure 5.11 can be used to compare mobility described by daily activity patterns at nodes with similar functions. For example, the patterns taken by nodes 1V and 2G, which are the exit nodes nearest to the subway station in their respective estates, can be compared. In the Group 4 column of Figure 5.11, node 1V is highly linked to cluster 4(vi) on Workdays, which has a pattern of morning and afternoon peaks of about with similar height, while node 2G is more highly linked to cluster 4(iii) on Workdays, which has a morning peak as well, but a less prominent afternoon peak. The middle peak observed at node 1V could be due to its proximity to node 1R, which is a LF area for the estate and contains some coffee shops that may be visited more frequently during the lunch hour.

Another observation that can be made with the spatial plots in Figure 5.11 is the comparison of the playground nodes which are heavily featured in Groups 2 and 3. Both playgrounds that are involved are in Estate 2, but node 2F is more heavily featured in Group 2, while node 2I is more heavily featured in Group 3. The activity patterns at node 2F may reflect the morning crowd that passes through while

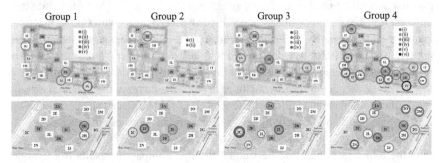

**FIGURE 5.11**   Maps showing locations of nodes that commonly appear in each of the clusters, for each of the groups. If a cluster is indicated in the legend but not displayed on the map, there is no node that commonly appears in that cluster.

*Source*: Authors.

heading for the nearby exit at node 2C, while parents may favour 2I to bring their children in the evenings after work as it is more central to the estate. This may help to inform design decisions about the layout of the facilities of the estate as well.

## Conclusion

In this chapter, a case study involving the use of passively sensed Wi-Fi data was presented. This case study involved data collected from two estates, and there were analyses performed on a single-estate level as well as an analysis over both estates.

For the single-estate data analysis, the data collected via passive Wi-Fi sensing was analysed in three ways. First, the daily temporal activity levels were visualised using line plots. These provided the insights that different DTs and node functions may influence daily activity patterns. For example, certain locations have morning peaks in activity levels only on working days and not on rest days such as weekends and public holidays. Next, linear regression analysis was performed on the data to identify how each factor of DT and rainy weather could affect the daily activity levels at nodes with different functions as compared to activity levels of the overall estate. Rainy weather tends to affect activity at nodes that are in the open, such as playgrounds and FCs, which suggests that some sheltered alternative facilities could be designed for children's activities during rainy weather. Lastly, HAC was performed to investigate how the residents moved about different parts within the estate. The grouped nodes showed some insights like a potential common transitory point within the estate, as well as certain closely linked points where residents passed by in both directions.

Additionally, a two-estate data analysis was performed by combining feature vectors extracted from data collected from two estates and clustering them to identify common temporal activity patterns from nodes with similar functions and on days of similar types. It was found that daily temporal patterns can be grouped into four main patterns depending on the number of peaks and the relative heights of the peaks at different times of day. A comparison was made between nodes with similar functions in both estates such as the node that is nearest to the exit to the subway station. These nodes in the two estates were associated with different clusters, indicating different activity patterns despite being similar in function, and there could be an influence in the activity levels due to nearby facilities as well. To increase the potential activity levels of certain planned facilities, these facilities could be placed nearer to, or on the way to, places with higher activity levels such as retail shops or coffee shops.

Apart from combining data from two estates and clustering their feature vectors, the two estates can also be studied through comparing results from single-estate data analysis on each estate. By comparing daily temporal activity levels, we can observe that the environmental deck (e-deck) at Estate 2 (Cascadia) is better used than the rooftop of multi-storey carpark at Estate 1 (Topaz) Multi-storey car park (MSCP). Yet, the ground-level nodes in Estate 1 show both higher and longer visitorship than nodes in Estate 2 on the e-deck. Based on visualisations of HAC,

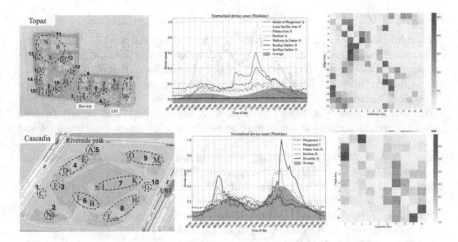

FIGURE 5.12  Comparison of data analysis between the two estates: (left) Daily temporal activity level of different spaces and nodes within each estate to show the usage of public space; (right) HAC for each estate to show the permeability and usage hotspots.

*Source*: Authors.

the e-deck at Estate 2 appears to facilitate more movements between nodes, in other words, it is more permeable than the ground-level space at Estate 1; but Estate 1 has more hotspots, i.e., attracts more people at its ground-level nodes, probably due to its proximity to surrounding commercial amenities. The question for community designers in this case would be, How might we attract more people to use the facilities or create more attractive nodes on e-deck while continuing to retain its permeability? This study has thus led to the subsequent intervention that the research team embarked on in Estate 2 (refer to Chapter 8).

In all, the insights drawn from this chapter have the potential to inform community design decisions, which include the locations and distributions of common facilities with respect to the planned entrances and exits to the estate, the impacts of different neighbourhood typologies such as e-deck and ground-level public space, as well as considering the potential influences of different types of facilities on the activity levels around each part of the estate.

## Acknowledgement

The authors would like to thank the Pasir Ris – Punggol Town Council for their support in the installation of sensors for the purpose of this study.

Part of this chapter is adapted from Zhou, Koh, Ng, Liu, Foo, Yuen, and Chong (2021). "Understanding Residents' Behaviours in a Housing Estate by Passive Wi-Fi Sensing and Data Mining". In *2021 Fifth World Conference on Smart Trends in Systems Security and Sustainability (WorldS4)*, pp. 45–51. IEEE.

## Bibliography

Alessandrini, A., Gioia, C., Sermi, F., Sofos, I., Tarchi, D., & Vespe, M. (2017). Wi-Fi positioning and big data to monitor flows of people on a wide scale. In *2017 European Navigation Conference (ENC)* (pp. 322–328). Lausanne, Switzerland: IEEE.

Bar-Joseph, Z., Gifford, D. K., & Jaakkola, T. S. (2001). Fast optimal leaf ordering for hierarchical clustering. *Bioinformatics, 17*(suppl 1), S22–S29.

Ciftler, B. S., Dikmese, S., Gümüş, İ., Akkaya, K., & Kadri, A. (2017). Occupancy counting with burst and intermittent signals in smart buildings. *IEEE Internet of Things Journal, 5*(2), 724–735.

Cleveland, W. S. (1979). Robust locally weighted regression and smoothing scatterplots. *Journal of the American Statistical Association, 74*(368), 829–836.

Comaniciu, D., & Meer, P. (2002). Mean shift: A robust approach toward feature space analysis. *IEEE Transactions on Pattern Analysis and Machine Intelligence, 24*(5), 603–619.

Davies, D. L., & Bouldin, D. W. (1979). A cluster separation measure. *IEEE Transactions on Pattern Analysis and Machine Intelligence, 1*(2), 224–227.

Depatla, S., & Mostofi, Y. (2019). Occupancy analytics in retail stores using wireless signals. In *2019 16th Annual IEEE International Conference on Sensing, Communication, and Networking (SECON)* (pp. 1–9). Boston, MA, USA: IEEE.

Ester, M., Kriegel, H.-P., Sander, J., Xu, X., et al. (1996). A density-based algorithm for discovering clusters in large spatial databases with noise. *Kdd, 96*(34), 226–231.

Graells-Garrido, E., Ferres, L., Caro, D., & Bravo, L. (2017). The effect of Pokémon Go on the pulse of the city: A natural experiment. *EPJ Data Science, 6*, 1–19.

Kalogianni, E., Sileryte, R., Lam, M., Zhou, K., Van der Ham, M., Van der Spek, S., & Verbree, E. (2015). Passive WiFi monitoring of the rhythm of the campus. In *Proceedings of the 18th AGILE International Conference on Geographic Informational Science* (pp. 9–16). AGILE.

Kleinbaum, D. G., Kupper, L. L., Nizam, A., & Rosenberg, E. S. (2013). *Applied regression analysis and other multivariable methods.* Duxbury: Cengage Learning.

Koh, Z., Zhou, Y., Lau, B. P. L., Yuen, C., Tuncer, B., & Chong, K. H. (2020). Multiple-perspective clustering of passive Wi-Fi sensing trajectory data. *IEEE Transactions on Big Data, 8*(5), 1312–1325.

Lloyd, S. (1982). Least squares quantization in PCM. *IEEE Transactions on Information Theory, 28*(2), 129–137.

MSS. (2021). *Historical daily weather records.* Retrieved from March 17, 2021, https://www.weather.gov.sg/climate-historical-daily/

Müllner, D. (2011). Modern hierarchical, agglomerative clustering algorithms. *arXiv preprint arXiv:1109.2378.*

Prentow, T. S., Ruiz-Ruiz, A. J., Blunck, H., Stisen, A., & Kjærgaard, M. B. (2015). Spatio-temporal facility utilisation analysis from exhaustive Wi-Fi monitoring. *Pervasive and Mobile Computing, 16*, 305–316.

Pu, Z., Zhu, M., Li, W., Cui, Z., Guo, X., & Wang, Y. (2020). Monitoring public transit ridership flow by passively sensing Wi-Fi and Bluetooth mobile devices. *IEEE Internet of Things Journal, 8*(1), 474–486.

Ribeiro, M., Nunes, N., Nisi, V., & Schöning, J. (2020). Passive Wi-Fi monitoring in the wild: A long-term study across multiple location typologies. *Personal and Ubiquitous Computing, 26*, 1–15.

Rousseeuw, P. J. (1987). Silhouettes: A graphical aid to the interpretation and validation of cluster analysis. *Journal of Computational and Applied Mathematics, 20*, 53–65.

Schauer, L., Werner, M., & Marcus, P. (2014). Estimating crowd densities and pedestrian flows using Wi-Fi and Bluetooth. *In Proceedings of the 11th International Conference on Mobile and Ubiquitous Systems: Computing, Networking and Services* (pp. 171–177).

Zakaria, C., Trivedi, A., Cecchet, E., Chee, M., Shenoy, P., & Balan, R. (2020). Analysing the impact of COVID-19 control policies on campus occupancy and mobility via passive Wi-Fi sensing. *arXiv preprint arXiv:2005.12050.*

Zhang, T., Ramakrishnan, R., & Livny, M. (1996). Birch: An efficient data clustering method for very large databases. *ACM SIGMOD Record, 25*(2), 103–114.

Zhou, Y., Koh, Z., Ng, B. K. K., Liu, R., Foo, S. X. D., Yuen, C., & Chong, K. H. (2021). Understanding residents' behaviours in a housing estate by passive Wi-Fi sensing and data mining. In *2021 Fifth World Conference on Smart Trends in Systems Security and Sustainability (WorldS4)* (pp. 45–51). London, UK: IEEE.

Zhou, Y., Lau, B. P. L., Koh, Z., Yuen, C., & Ng, B. K. K. (2020). Understanding crowd behaviours in a social event by passive Wi-Fi sensing and data mining. *IEEE Internet of Things Journal, 7*(5), 4442–4454.

# 6

# SOCIAL NETWORK ANALYSIS AS A DATA-LIGHT METHOD FOR IDENTIFYING PLACES OF SOCIAL INTERACTION

*Zi-Yu Khoo, Aerilynn Tan and Keng Hua Chong*

## Introduction

As of 2018, over half of the world's population resides in urban areas (United Nations, 2018). For urban residents, public spaces provide space for social interaction and activities. Public spaces that offer entertainment, provide amenities that bring comfort, have seating spaces, or offer shelter are also attractors for residents (Holland, Clark, Katz, & Peace, 2007), increasing the opportunity for social interaction.

However, not all public spaces become venues for social interaction. The identification of public spaces that attract residents and encourage social interaction has become of interest – due to the COVID-19 pandemic, individuals have been advised to avoid large gatherings. Singapore's Urban Redevelopment Authority (URA) has introduced crowd monitoring and diversion measures to reduce crowding at public spaces (Gov.sg, 2020). The whole of government has also deployed Safe Distancing Ambassadors to physically patrol and verbally remind Singaporeans to practise safe distancing at public spaces (Singapore Government Developer Portal, 2020). However, while crowd levels at commercial spaces are manually monitored by their operators, non-commercial spaces are not, and are too numerous to station staff at, to monitor continuously. Therefore, there is a need to distinguish public spaces that are more attractive to residents for social interaction to efficiently distribute resources for safe distancing management.

This chapter introduces the methodology and validation of a study that identifies public spaces which are likely to play host to social interaction. It is contextualised to a selected residential town in Singapore, but its application is generalisable. Public spaces in this study refer to precinct facilities provided in Singapore's public housing developments, catering to the social and recreational needs of the

DOI: 10.4324/9781003437659-9

immediate residents (Housing & Development Board [HDB], 2005). The study creates a network of precinct facilities in the selected residential town, and performs a Social Network Analysis. Based on degree centrality measures, geographic location, and type of each precinct facility, points with high opportunity for social interaction are found and validated through comparison with field observations. Additional observations are also made regarding how precinct facilities are used, designed, and distributed, which enables social interaction. With these observations, resources for safe distancing management can be effectively distributed among these precinct facilities to prevent the spread of COVID-19.

## Case Study

### Study Site

The study site is Toa Payoh, Singapore. The town is currently home to 121,700 residents. Toa Payoh was originally prioritised for massive resettlement. Over the years, other goals were introduced, including to nurture community spirit and social interactions (Cheong, 2017). Insight into the location of social interactions can help policy makers deploy staff for safe distancing management, and urban planners to modify existing towns by introducing or removing precinct facilities.

### Social Interaction

Social networks and human interaction can be understood using social media data found online or via Wi-Fi, location-based services, and the Internet of Things. For social media, networks created using follower data between Twitter users identify the decline of network density with increasing geographical distance (Stephens & Poorthuis, 2015) that suggest prioritisation of edges connecting nodes near each other. Furthermore, formed communities are centred around common interests of users suggesting that similar physical communities can be formed around precinct facilities that support shared common interests of residents (Van Meeteren et al, 2010), such as chess-playing or exercise.

Pedestrians can also be accurately localised using mobile phones in indoor environments with Wi-Fi and pedestrian dead reckoning (Beom et al, 2010) and outdoor environments with sensors and GPS information (Wu et al, 2015; Yuan & Chen, 2014). However, residents such as young children and senior citizens may not carry mobile phones. This motivates research for a data-light method to localise residents to precinct facilities where social interaction can take place.

### Social Network Analysis

Social Network Analysis is a methodology for understanding and investigating social structures. It considers units in a network and exemplifies their relationships. Analytical methods applied to the social structure use the network structure to deal directly with the relationships between units instead of the attributes of each individual unit (Wellman, 1988). Following the growing use of Social Network Analysis to understand the relationship between different units such as transport nodes (Rodrigue, 2020), individuals (Sola Pool & Kochen, 1978), and buildings (Chowell et al, 2003), this methodology can be applied to precinct facilities, to understand their relationships and relative importance to residents.

Degree centrality represents the involvement of each node in a network, quantified through the weights of adjacent edges of each node. In searching for hubs within the human brain network, the variation in degree centrality across a network identifies the relative importance of points in the network. A well-defined network can provide an understanding of the potential influence of highly connected node areas and subsequently be used to find convergence points (Fornito et al, 2016). This methodology can be applied to a network of precinct facilities, to identify convergence points for social interaction.

## Data

Locations for precinct facilities were found from Data.Gov.Sg, or through field observations from Toa Payoh. For all types of precinct facilities, refer to Figure 3.3.

The study site included all residential areas on roads named Toa Payoh. The site was divided into verification and validation sites. The smaller verification site was used to build and tune the model to identify locations for social interaction. The model was then applied to the larger validation site. The study site with precinct facilities can be observed in Figure 6.1.

## Types of Social Interaction in Relation to the Environment

Four types of social interaction in precinct facilities were of interest, categorised based on how residents use and interact within them. These categories include passive, reactive, and creative (Hill, 2001), with a fourth type, 'active', proposed based on field observations.

### Passive Usage

Passive usage of precinct facilities refers to when residents spontaneously make use of facilities for their intended purpose, "transforming neither use, space nor meaning" (Hill, 2001). Figure 6.2 shows children playing together in a playground as an example of passive usage. Social interaction of observational interest involved residents at the facility that interacted spontaneously through conversation or participation in the same activity. The social interaction that took place was a product of the precinct facility.

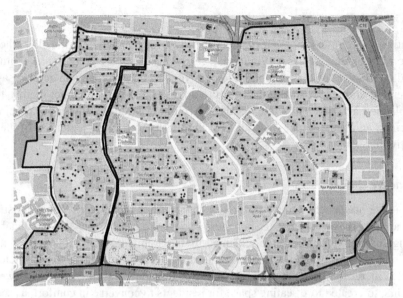

FIGURE 6.1    Boundaries and distribution of precinct facilities in the Toa Payoh verification (left) and validation (right) sites, superposed on the OpenStreetMap layer available in QGIS 3.4.10; 697 precinct facilities were identified, comprising 219 in the verification site and 468 in the validation site.

*Source*: Authors.

FIGURE 6.2    Top left: Children from different families spontaneously playing together through passive usage of the playground. Top right: Active use of sheltered pavilion beside the RC at Block 141 Lorong 2 Toa Payoh, by organising Silat classes. Bottom left: Reactive usage of void deck space, by adding chairs to existing benches to facilitate conversations. Bottom right: Creative appropriation of green buffer space in front of ground-floor units.

*Source*: Authors.

### Active Usage

Active usage refers to using precinct facilities for planned activities like chess games or exercise classes (Figure 6.2, top right). They are consistent in nature and of a larger scale as compared to passive usage. Participants in active usage make separate independent decisions to arrive at a specific precinct facility for a planned activity. They interact due to common interests within the precinct facility, and usually take on a sense of familiarity due to the regularity of these activities.

### Reactive Usage

Reactive usage refers to when a resident "modifies the physical characteristics of a space as needs change, but must choose from a narrow and predictable range of configurations largely defined by the architect" (Hill, 2001). Such additions become pseudo-permanent, and are added reactionarily to insufficient infrastructure at precinct facilities. Most of the added furniture observed were chairs, to create more seating space for residents to converse in comfort, as seen in Figure 6.2, bottom left. This temporal or semi-permanent modification of precinct facilities also allowed social interaction to be identified even when unused during observation.

### Creative Usage

Creative users "either create a new space or give an existing one new meanings and uses" (Hill, 2001). In the context of Singapore's public housing, these usually involve the appropriation of unused or underutilised "white spaces", usually initiated by an individual or a group of like-minded residents who share a common desire to inject a sense of identity and a new point of interest in their neighbourhood (Chong & Cho, 2018). For example, residents were observed to appropriate spaces along empty green buffers outside ground-floor housing units with pop-up gardens and outdoor furniture. An example of creative usage can be seen in Figure 6.2, bottom right. Social interaction occurs when neighbours visit these spaces, whether out of curiosity, by coincidence, upon invitation, or self-invited. Mutual trust is required between the residents for its success, due to risks of vandalism and theft.

### Combination of Usage

Combination-type usage for social interaction was also observed. This was observed at Block 179 Toa Payoh Central, where residents were observed to make passive use of public chess tables to play chess, but also reactively left their personal chairs near the chess table to sit while watching games. This constituted both reactive and passive usage of the precinct facility for social interaction.

## Methodology

### Social Network Creation

A simple undirected spatial network was created using each precinct facility as a node. The geographic location of all nodes was identified and the straight-line distance between them calculated. Edges in the spatial network were weighted such that spatial interaction between two nodes is inversely related to the square of their distance apart (Pun-Cheng, 2016):

$$Edge\ weight = \frac{1}{\left(Distance\ Apart\right)^2} \qquad (1)$$

For a weighted network, the degree centrality of a node is calculated by aggregating weights of all its edges, where $C_D^W(i)$ is the weighted degree centrality of node $i$, and $w$ is the weight of the edge connecting nodes $i$ and $j$.

$$C_D^W(i) = \sum_{j \neq i}^{N} W_{i,j} \qquad (2)$$

### Tuning via Application to Verification Site

Within the verification site, nodes with a comparatively higher degree centrality than other nodes within a search radius were identified. A search radius of 100 m was ideal – it was fine enough to identify local peaks in degree centrality within the network, and coarse enough to capture a sufficient number of nodes. The results were consistent and stable when the search radius varied between 85 and 115 m.

Of the 20 nodes identified in the verification site, 8 corresponded to commercial or social amenities. As these are programmed spaces with a pre-specified purpose, they lack space for social interaction, and were replaced with the closest distance non-programmed space.

Using this replacement methodology, 20 nodes representing non-programmed precinct facilities were identified for potential social interaction. A post-analysis visit to the verification site found that 17 out of the 20 identified nodes hosted social interaction at the precinct facility itself, or directly adjacent to it.

## Results and Analysis

A network was built using 468 precinct facilities within the validation site. Identical to the methodology applied to the verification site, nodes of high degree centrality were found using a 100 m search radius, then filtered to replace programmed spaces.

Combining the verification and validation sites, 57 nodes with high potential for social interaction were identified. Figure 6.3 gives a breakdown of their types

TYPES AND NUMBER OF EACH PUBLIC SPACE

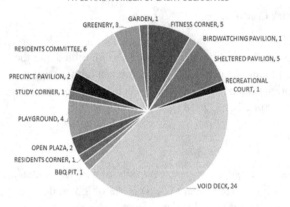

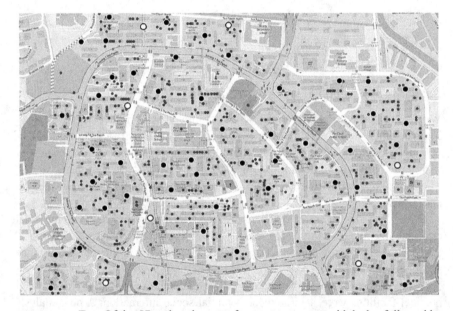

**FIGURE 6.3** Top: Of the 57 nodes, the most frequent type was void decks, followed by RCs, fitness corners, sheltered pavilions, and playgrounds. Bottom: Nodes of high degree centrality where social interaction was found (large black points), or not found (large white points), superposed on 697 precinct facilities.

*Source*: Authors.

and locations. Social interaction was found through field observations at 50 of the 57 nodes. Details regarding the social interaction and passive, active, and reactive usage at these 50 nodes can be found in Figures 6.4–6.6.

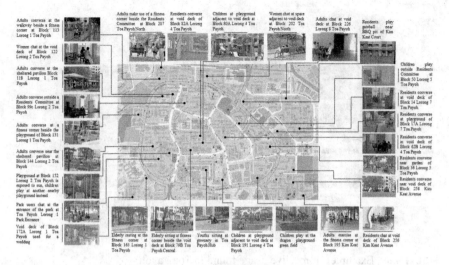

**FIGURE 6.4**   Examples of passive usage found in Toa Payoh.

*Source*: Authors.

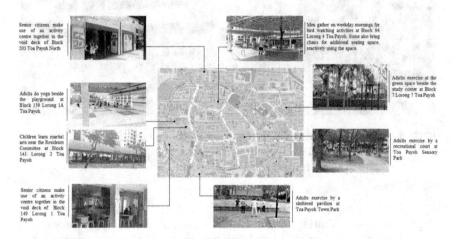

**FIGURE 6.5**   Examples of active usage found in Toa Payoh.

*Source*: Authors.

## Discussion

### *Void Decks*

Void decks refer to the ground floor of public housing – a design decision of the Housing & Development Board to construct public housing on "pillars" and free up ground-level open, sheltered spaces (Tan et al, 2013); 23 of the 24 void decks identified had a high degree of centrality because they shared space with other

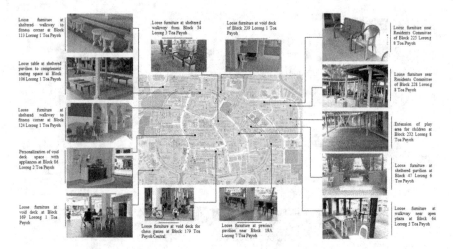

**FIGURE 6.6** Examples of reactive usage found in Toa Payoh.

*Source:* Authors.

**FIGURE 6.7** Left: Residents having spontaneous conversation at void deck in Block 122 Lorong 2 Toa Payoh. Right: Block 202 Toa Payoh North, residents chat at an open space adjacent to the void deck.

*Source:* Authors.

programmed spaces. During field observations, 20 of the 24 void decks hosted social interaction with passive usage by residents who made use of programmed spaces in the void deck, were waiting for the programmed spaces to open, or passed through the void deck to reach other precinct facilities, as demonstrated in Figure 6.7.

In some cases, void decks were close to other precinct facilities but lacked space for activities, resulting in residents overflowing to adjacent spaces, or reactively bringing their own chairs, as shown in Figure 6.7.

### Residents' Committees

Six out of 20 Residents' Committees (RCs) in Toa Payoh were found to have high social interaction potential. All six hosted social interaction. Five shared a void deck space with other programmed spaces and three were adjacent to a playground or fitness corner. Another seven RCs in Toa Payoh were near other points of high potential for social interaction. RCs foster harmonious inter-ethnic relationships through their activities (Sim et al, 2003), and encourage interaction by providing seating spaces for passive usage and spontaneous conversation near these facilities. Social interaction near RCs are seen in Figure 6.8.

RCs are also close to clusters of non-programmed spaces, have high degree centrality, and can attract residents to attend organised events. These promote active usage of precinct facilities shown in Figure 6.8, enabling social interaction.

### Sheltered Pavilions

Five sheltered pavilions were identified to have high social interaction potential. All hosted social interaction due to their proximity to other programmed spaces, playgrounds, and fitness corners. An example is shown in Figure 6.9, left.

**FIGURE 6.8**    Left: Residents at Block 207 Toa Payoh North chatting and making passive use of fitness corners behind a RC. Right: Residents at Block 228 Lorong 8 Toa Payoh, chatting and making passive use of fitness corners behind a RC.

*Source*: Authors.

**FIGURE 6.9**    Left: Residents make active use of a sheltered pavilion for a yoga session by a playground at Block 138C Lorong 1A Toa Payoh. Right: Residents (parents) make use of a fitness corner, and chat, while their children play in a playground at Block 131 Toa Payoh Crest.

*Source*: Authors.

## Fitness Corners and Playgrounds

Five fitness corners and four playgrounds were found to have high social interaction potential. Of these, eight hosted social interaction – six were clustered between programmed spaces while two were between precinct pavilions or void decks, and playgrounds, resulting in their high potential for social interaction. Most involved passive usage of the precinct facility.

Forty-eight out of 60 playgrounds (80%) and 55 out of 72 fitness corners (76.3%) were built within 100 m of another fitness corner or playground. This pairing of playgrounds and fitness corners resulted in their high and similar degree centralities. This resulted in children playing with the exercise equipment at fitness corners. Simultaneously, their elderly caretakers were able to use the fitness corners, facilitating inter- and intra-generational social interaction, such as in Figure 6.9, right.

## Other Social Interaction

The Social Network Analysis failed to identify social interaction at a sheltered walkway outside a RC at Block 29 Lorong 5 Toa Payoh. Its closest node of high social interaction potential was a nearby playground, with degree centrality 1.3% higher than that of the RC. Reactive use of this pedestrian linkway provided sheltered seating for residents visiting the RC or playground nearby, supporting the idea of RCs building on existing social interaction or attracting residents for social interaction.

## Locations Lacking Social Interaction

Seven out of 57 nodes of high degree centrality did not host social interaction. Of these, three were observed to have heavy human traffic, enabling spontaneous conversations. The remaining four were under 150 m from other points of high centrality with very heavy usage, which may have diverted residents.

## Conclusion

The proposed Social Network Analysis is an effective and data-light methodology for identifying social interaction in precinct facilities. Of the 57 nodes recommended by the analysis, 50 were found to host social interaction through field observations. This analysis can hence be used by urban planners as a data-light method to understand how planning or placement of precinct facilities can impact the development of social dynamics and relations; or by urban ecologists looking for study sites to understand social interaction.

A possible improvement of this study is to make use of degree centrality and the type of precinct facility as inputs to identify the type of social interaction that would take place or how the precinct facility would be used. This can potentially be done through unsupervised machine learning techniques, if there would be sufficiently large dataset.

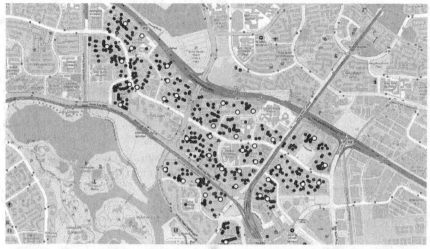

**FIGURE 6.10** Top: Of the 41 nodes of high centrality predicted by the model, 26 were found to have social interaction (large white dots), from a total of 534 precinct facilities in Jurong East. Bottom: 47 nodes were predicted by the model to have high propensity for hosting social interaction, from a total of 810 precinct facilities in Punggol. However, the research team did not manage to carry out site verification due to the COVID-19 restrictions at the time of research.

*Source*: Authors.

This analysis had been applied to the other two residential towns, Jurong East and Punggol, and had helped the research team in identifying intervention sites in these towns (Figure 6.10). This chapter has highlighted the importance of the placement of precinct facilities, and that the data-light analytical method could serve to provide quick insights to the social interaction of residents.

## Acknowledgement

The Social Network Analysis method developed in this research has been integrated into the City Application Visualisation Interface (CAVI) platform developed by Électricité de France, EDF Lab Singapore. Through the CAVI platform, users can dynamically add and remove precinct facilities, and make use of the in-built Social Network Analysis to understand how their changes made to precinct facilities can affect social interaction in the vicinity. The authors are grateful to EDF Lab Singapore for supporting this research.

The earlier version of this chapter was published in a conference proceeding of the Cities in a Changing World Conference 2021: Khoo et al (2021).

## Bibliography

Beom, J. S. et al. (2010). Indoor wifi positioning system for Android-based smartphone. *2010 International Conference on Information and Communication Technology Convergence, ICTC 2010* (pp. 319–320). https://doi.org/10.1109/ICTC.2010.5674691

Centres for Disease Control and Prevention. (2019). *Considerations for events and gatherings*.

Chen, Z., Zou, H., Jiang, H., Zhu, Q., Soh, Y. C., & Xie, L. (2015). Fusion of WiFi, smartphone sensors and landmarks using the Kalman filter for indoor localization. *Sensors (Switzerland), 15*(1), 715–732. https://doi.org/10.3390/s150100715

Cheong, K. H. (2017). The evolution of HDB towns. In C. K. Heng (Ed.), *50 Years of urban planning in Singapore* (pp. 101–125). Singapore: World Scientific.

Chong, K. H., & Cho, M. (2018). *Creative ageing cities: Place design with older people in Asian cities*. London: Routledge.

Chowell, G., Hyman, J. M., Eubank, S., & Castillo-Chavez, C. (2003). Scaling laws for the movement of people between locations in a large city. *Physical Review E – Statistical Physics, Plasmas, Fluids, and Related Interdisciplinary Topics, 68*(6), 066102. https://doi.org/10.1103/PhysRevE.68.066102

Fornito, A., Zalesky, A., & Bullmore, E. T. (2016). Node degree and strength. In *Fundamentals of brain network analysis* (pp. 115–136). Elsevier. https://doi.org/10.1016/b978-0-12-407908-3.00004-2

Gov.sg (2020, Jun 10). *Keeping you safe by keeping you apart*. Retrieved 2021, June 11, from https://www.gov.sg/article/keeping-you-safe-by-keeping-you-apart

Hill, J. (2001). The use of architects. *Urban Studies, 38*(2), 351–365. https://doi.org/10.1080/00420980123765

Holland, C., Clark, A., Katz, J., & Peace, S. (2007). *Social interactions in urban public places*. Bristol: The Policy Press.

Housing & Development Board (HDB). (2005). *Public housing design guide: Principles and practice*. Singapore: Housing & Development Board (HDB).

Khoo, Z.-Y., Tan, A., & Chong, K. H. (2021). To congregate or to disperse? Structural analysis of public spaces for social interaction identification. In J. Montgomery

(Ed.), AMPS proceedings series 24.1. Cities in a changing world (pp. 38–51). New York: City Tech – CUNY.

Liu, T. K. (2017). Planning and urbanisation in Singapore. In C. K. Heng (Ed.), *50 Years of urban planning in Singapore* (pp. 23–44). Singapore: World Scientific.

Otte, E., & Rousseau, R. (2002). Social network analysis: A powerful strategy, also for the information sciences. *Journal of Information Science, 28*(6), 441–453. https://doi.org/10.1177/016555150202800601

Population Division of the Department of Economic and Social Affairs of the United Nations. (2018). *World urbanization prospects*. Retrieved from https://population.un.org/wup/

Pun-Cheng, L. S. C. (2016). Distance decay. In *International encyclopedia of geography: People, the earth, environment and technology* (pp. 1–5). John Wiley & Sons, Ltd. https://doi.org/10.1002/9781118786352.wbieg0179

Rodrigue, J.-P. (2020). *The geography of transport systems* (5th ed). London: Routledge.

Shin, B. J., Lee, K. W., Choi, S. H., Kim, J. Y., Lee, W. J., & Kim, H. S. (2010). Indoor WiFi positioning system for Android-based smartphone. In *2010 International Conference on Information and Communication Technology Convergence, ICTC 2010* (pp. 319–320). https://doi.org/10.1109/ICTC.2010.5674691

Sim, L. L., Yu, S. M., & Han, S. S. (2003). Public housing and ethnic integration in Singapore. *Habitat International, 27*(2), 293–307. https://doi.org/10.1016/S0197-3975(02)00050-4

Singapore Government Developer Portal. (2020). *Space out - Crowd analysis for safe distancing*. Retrieved 2021, June 11, from https://www.developer.tech.gov.sg/technologies/digital-solutions-to-address-covid-19/space-out

Sola Pool, I. de, & Kochen, M. (1978). Contacts and influence. *Social Networks, 1*(1), 5–51. https://doi.org/10.1016/0378-8733(78)90011-4

Stephens, M., & Poorthuis, A. (2015). Follow thy neighbour: Connecting the social and the spatial networks on Twitter. *Computers, Environment and Urban Systems, 53*, 87–95. https://doi.org/10.1016/j.compenvurbsys.2014.07.002

Tan, A., Sim, C., & Tan, B. (2013). Void decks community heritage series III. Singapore: National Heritage Board.

Van Meeteren, M., Poorthuis, A., & Dugundji, E. (2010). Mapping communities in large virtual social networks: Using Twitter data to find the Indie Mac community. In *2010 IEEE International Workshop on Business Applications of Social Network Analysis, BASNA 2010*. https://doi.org/10.1109/BASNA.2010.5730297

Wellman, B. (1988). Structural analysis: From method and metaphor to theory and substance. In B. Wellman & S. D. Berkowitz (Hrsg.), *Social structures: A network approach* (pp. 19–61). Cambridge: Cambridge University Press.

Wu, C., Yang, Z., Xu, Y., Zhao, Y., & Liu, Y. (2015). Human mobility enhances global positioning accuracy for mobile phone localization. *IEEE Transactions on Parallel and Distributed Systems, 26*(1), 131–141. https://doi.org/10.1109/TPDS.2014.2308225

Yuan, Q., & Chen, I. M. (2014). Localization and velocity tracking of human via 3 IMU sensors. *Sensors and Actuators, A: Physical, 212*, 25–33. https://doi.org/10.1016/j.sna.2014.03.004

# 7

# PREDICTIVE ANALYSIS WITH VIRTUAL POPULATION

*Roland Bouffanais, Malik Mohamed Barakathullah and Bige Tunçer*

## Introduction

The evolution of population demographics over time and space is a phenomenon that warrants deeper comprehension, especially in areas like urban planning and infectious disease prevention.

Indeed, shifts in demographics over time within a population are influenced by key events in an individual's life. Factors like birth, death, immigration, and emigration play roles in the overall population changes in a nation. However, internal movements, motivated by the pursuit of amenities and resources essential for daily life, drive the spatial dynamics.

The rates of such events are influenced by several factors, including education, income, access to healthcare and other amenities, overall quality of life, internal or neighbouring conflicts, ecological conditions, and more. If the rates of these events remain relatively consistent over time, modelling the temporal demographic shifts becomes straightforward.

The population's changes due to these events can be modelled using a dynamical-system approach, focusing on the sizes of distinct population groups, or via agent-based modelling (ABM), where each individual's progress and interactions reflect the probable events in their life, as informed by census data. Such models equip policymakers and planners with tools to envision multiple scenarios before making decisions.

In Singapore, public housing developed by the Housing & Development Board (HDB) is home to over 80% of its residents. Accurate demographic modelling can assist HDB in planning based on diverse population structures, including age, household makeup, and family configurations. Unique to Singapore, HDB estates implement ethnic quotas, ensuring an even spatial distribution of various ethnic groups throughout the city-state.

DOI: 10.4324/9781003437659-10

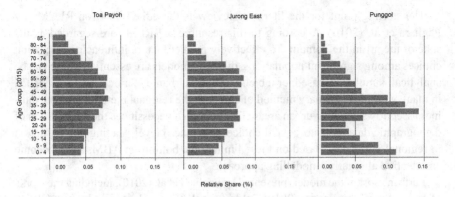

**FIGURE 7.1**    Age distribution (in 2015) of considered types of the towns.

*Source*: Department of Statistics, Singapore.

Anticipating demographic shifts in each town's public housing estates is essential for timely planning of necessary amenities. For instance, if anticipated birth rates in a neighbourhood are expected to surpass historical figures, there might be a future need for additional primary schools.

In Singapore, towns exhibit distinct demographic patterns based on their historical origins and developmental timelines. For instance, older towns like Toa Payoh predominantly house an ageing population, resulting in a skewed age distribution, as depicted in Figure 7.1 (left). Conversely, 'newer' towns like Punggol often attract younger couples and exhibit a dual-peak age distribution, evident in Figure 7.1 (right). Lastly, Jurong East shows a relatively even age distribution between 20 and 60 years due to continuous development phases, as illustrated in Figure 7.1 (middle).

Each town is made up of multiple neighbourhoods, which are further divided into housing precincts containing numerous apartment blocks. The demographic shifts within a neighbourhood largely hinge on the recency of its estate developments and the family compositions residing within. Thus, to accurately forecast demographic changes at a neighbourhood level, one cannot solely depend on aggregated national statistics. A detailed survey is imperative to ascertain the family dynamics and age profiles specific to each neighbourhood. Furthermore, any attempt to forecast the move from one neighbourhood to another requires the development of a residential location choice model (RLCM).

## Review of RLCMs: Synthetic Population, Population Dynamics, and Location Choice Models

In this section, we review the three key components of agent-based residential location models (ABRLMs), namely, the initial population generation, the model for demographic changes, and the location choice model.

The starting point for the literature review is the edited book on RLCMs by Pagliara et al. (2010). According to this resource, ABRLMs are regarded as the state of the art in the domain. To effectively predict factors influencing relocation choices among significant populations, dynamic models are essential. These can be analytical, simulation-based, or a blend of both. Urban dynamics can be modelled in many ways. The primary method of choice is the regional method, which examines asset flows between urban areas. This approach necessitates the aggregation of demographic data and subsequent choices at a macro level, leading to a reduction in regional variances. Based on these limitations, bottom-up ABMs have become an effective alternative modelling approach.

Further, most of the models presented by Pagliara et al. (2010), including the works of Arentze and Timmermans (2010), Feldman et al. (2010), Hunt and Abraham (2010), and Waddell (2010), are part of larger transport-land-use interactions models, whereby the interplay between transport accessibility and land use, residential and commercial, are modelled. In fact, most of the advances in RLCMs are made within such larger modelling frameworks (Acheampong & Silva, 2015; Moeckel et al., 2018).

Moeckel (2017), in developing the integrated land-use agent-based model SILO (Simple Integrated Land Use Orchestrator), utilises five primary components depicted in Figure 7.2. Out of these, three can be regarded as riendational elements for ABRLMs. Initially, a synthetic household population representing agents and their attributes is produced. Following this, changes in household demographics are projected over time, leading to shifts in household members and the overall household structure. Subsequently, household relocations occur based on environmental factors and demographic changes. These three components are briefly reviewed in the following subsections.

### Agent Attributes and Population Synthesis

In deploying an agent-based microsimulation model like an ABRLM, the initial task involves determining and constructing agents that make up the synthetic population along with their associations (Müller & Axhausen, 2010). In the context of ABRLMs, agents are symbolic of individuals residing within the target area, and they are organised into households. These households are then endowed with attributes that influence their relocation decisions. Rather than predetermining these attributes, modellers typically base them on the data pertinent to the study (Brown & Robinson, 2006; Heldt et al., 2016). Surveys can also be conducted to acquire fresh data for the investigation, as exemplified by Buchmann et al. (2016) and Van Eggermond et al. (2018). Given that present-day housing features play a role in deducing household preferences, these attributes are directly linked to the household. As the model progresses, alterations in household and dwelling attributes transpire based on predefined guidelines and the actions of agents.

The goal of attributing specific characteristics to households is to create diverse households for the simulation. Households predicted to act similarly based

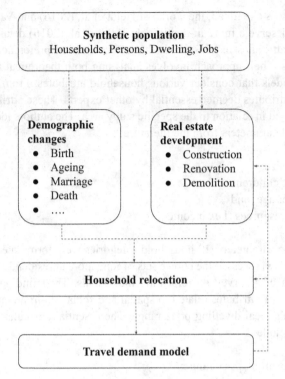

**FIGURE 7.2**   Five key model components of ABRLMs.

*Source*: Moeckel (2017).

on their features are given identical decision-making guidelines. Additionally, in a comprehensive model, the attributes and behaviour of households can evolve over time and through interactions with other agents and their surroundings (Huang et al., 2013). The significance of precise agent diversity was examined by Buchmann et al. (2016). They determined that this diversity is most influential when precision is needed on smaller scales, like at the precinct level. Hence, household diversity is vital for the New Urban Kampung (NUK) research programme.

Six primary scholarly sources were referenced to pinpoint common attributes assigned to households in ABRLMs. These are Brown and Robinson (2006), Buchmann et al. (2016), Heldt et al. (2016), Huang et al. (2014), Schirmer et al. (2014), and Van Eggermond et al. (2018). Some attributes are specific to certain locations and might be viewed as environmental factors. However, they remain closely tied to households, informing relocation preferences. While it would be ideal to assign all potential attributes to households in a model, practicality dictates that attributes come from existing data sources, sometimes supplemented by select survey data. +++Thus, the model is crafted based on these available traits. It is beneficial to focus on a select few attributes for prioritisation.

Of the sources examined, the works of Heldt et al. (2016) and Van Eggermond et al. (2018) deserve a more in-depth look. Heldt et al. (2016) detail a method to pinpoint key attributes in households that affect location preferences in location choice models. Their approach involves analysing both theoretical and empirical studies on models that consider various household attributes. From this analysis, they extract attributes deemed essential by other experts. These attributes are then further evaluated in relation to the specific study area. The authors identify the following critical parameters from the literature:

a   household size;
b   presence of children;
c   householder age; and
d   household resources (i.e., income).

Based on these parameters, 12 household categories were formulated. Van Eggermond et al. (2018) assessed the choice sets of Singaporean residents, drawing from empirical data of a survey on relocating individuals. Their findings highlighted the importance of attributes related to spatial and social environments, as well as the household's ideal dwelling price range. The essential attributes identified for households include:

a   preferred dwelling price range;
b   average distance to work;
c   distance to parents/relatives;
d   average distance where they most frequently meet their five closest contacts;
e   distance to a top primary school; and
f   proximity to mass rapid transit.

The authors suggest that, due to the significance of spatial variables, future studies should explore and analyse other anchor and distance-related attributes not covered in their research. Given the pertinence of these two publications to the NUK programme, the aforementioned critical attributes were prioritised.

After defining the agents, the subsequent step involves creating a collective of these agents. The optimal scenario would be to obtain a precise snapshot of the current HDB population, detailing their essential attributes for crafting an ABRLM. However, due to issues like privacy and associated costs, such data is often elusive. To address this, scholars have adeptly merged varied data sources through population synthesis. This method yields a detailed, diverse representation of agents that aligns with provided aggregate data (Müller & Axhausen, 2010). This collective of agents and households is termed the synthetic population.

Two such populations have already been developed for Singapore in other agent-based models. The first, MATSim Singapore, is a comprehensive agent-based transport simulation that accommodates agents' daily travel patterns based on individual attributes. Sun and Erath (2015), and Sun et al. (2018) have detailed the use of an intricate synthetic population in this platform. The foundational data for this population is derived from the 2012 Household Interview Travel Survey (HITS) conducted by Singapore's Land Transit Authority, coupled with publicly accessible census data. Many of the previously mentioned essential attributes are part of this population. Conversely, SimMobility offers a long-term planning module, emphasising both residential and commercial location choice modelling, aligning with the needs of the NUK project. For this platform, Zhu and Ferreira Jr (2014) crafted a synthetic household population grounded in the 2008 HITS survey, 2012 census data, and augmented with commercial datasets and insights from HDB and REALIS.

For the NUK programme, we leveraged the pre-existing synthetic populations rather than constructed an entirely new one. These established populations could then be enhanced by adding essential attributes relevant for ABRLMs.

## Demographic Changes

After establishing the initial population, the subsequent phase in the model involves simulating alterations in household composition over time. Events like marriage, childbirth, divorce, and retirement can serve as catalysts prompting residents to consider relocation.

Moeckel (2017) models demographic changes in SILO using Markov models by applying transition probabilities. Eight possible transitions are considered, namely:

a Ageing;
b Birth of a child;
c Leave parental household;
d Get married/cohabitate;
e Get divorced/separate;
f Change job;
g Variation in income;
h Death.

The likelihood of any of these events occurring to an agent is determined based on the agent's attributes, the household's attributes, and public statistical data. During each simulation phase, these events are randomly sequenced across all residences. Some events may necessitate an immediate move, while others may elevate the chances of relocation. A notable constraint of SILO is the static nature of these probabilities over time. Factors like birth rates for specific demographic groups or life expectancies remain unaltered in the projected years.

Buchmann et al. (2016) outline specific triggers for a data-driven ABRLM. While they employ triggers akin to those in SILO, they don't uniformly rely on probability. For example, when a household agent reaches 18 years, they depart from the household. Trigger probabilities are rooted in an extensive mover survey tailored to their research area. Triggers and events are activated up to a predetermined, year-specific threshold. These rates are derived from actual demographic data spanning 2001–2012. The chance of an event impacting an agent or household fluctuates based on household types and the rich empirical data at hand. The model for demographic shifts can also function independently and then interface with the relocation module.

UrbanSim, as described by Waddell (2010), adopts this methodology. There are various tools created for this purpose, with Geard and colleagues' 2013 model being noteworthy. Their tool projects household formations and evolutions over a defined period. Fundamental life milestones like birth, death, marriage, divorce, and childbirth are incorporated. An advantage of their model is its straightforwardness, which simplifies its implementation and maintenance. The model is also adept at crafting an initial synthetic population, subsequently modelling its evolution while monitoring household structure and makeup.

For the NUK programme, the streamlined methods of Moeckel (2017) or Geard and colleagues (2013) were adopted by tailoring them to Singapore's national demographic data. Should detailed residential information from HDB become accessible, we could have refined the rates to enhance the model's precision.

### Household Relocation

As household compositions evolve over time, some are prompted to move. Meanwhile, some households leave the system while new ones join. The departure of households frees up apartments, leading to competition among incoming and relocating households for these available spaces.

Many ABRLMs utilise logit models to determine potential relocation destinations for households. Waddell's 2010 UrbanSim model employs a Multinomial Logit Model, where relocating households assess available residential properties based on their features and price, juxtaposed with their own attributes, such as income and household size. Utility functions, combining both household and property attributes, gauge the appeal of each property. This utility then shapes the likelihood of a household selecting a particular location. Various strategies dictate the allocation of properties to agents, one of which is granting the property to the household that derives the highest utility from it. This assumes that such households would be more inclined to offer a higher price, thus outbidding others. Moeckel (2017) also integrates constraints into household choices, emphasising that certain attributes are non-negotiable for households. After relocations, demographic adjustments are made, triggers are applied, and the cycle persists. The model concludes after a predetermined time span.

A notable amount of research has been undertaken to effectively capture household preferences, as evident in platforms like UrbanSim and SILO. For the NUK programme, we sought to leverage these existing models. Nonetheless, adaptations were essential to accommodate the specific HDB guidelines and standards in practice.

This section reviews a specific range of ABRLM features that have been integrated within our models. For this endeavour, we established a model akin to Waddell's (2010) RLCMs, combined with the demographic shift model by Geard and colleagues (2013). This synthesis incorporated the trio of primary components discussed earlier. The goal was to adapt existing datasets and models to the specifics of the NUK programme, thus bypassing the need for building models from the ground up. The following section delves into an analysis of current model implementations, from which the optimal model for application in the NUK programme was selected.

### Review of Existing Implementations: Available Population Dynamics and Location Choice Software

In the NUK programme, an ABRLM has been implemented to predict how individual and household compositions would change over time, and how households would relocate. To reduce the development time of the model, existing applications were reviewed with the objective of identifying and repurposing the best-suited applications. Three promising applications were identified, of which UrbanSim (Waddell, 2010) to be linked with the Sim-Demog (Geard et al., 2013) population dynamics model was ultimately chosen. This section outlines the selection process followed.

We started our exploration of model applications by consulting the authoritative book on RLCMs authored by Pagliara et al. (2010). This seminal work, the singular comprehensive reference on RLCMs, delves into their development within the context of intricate land-use-transport-interaction models. From this foundational text, we discerned pertinent submodels applicable to our project. Subsequently, our inquiry led us to four pivotal review articles in this domain: Acheampong and Silva (2015), Hunt et al. (2005), Kii et al. (2016), and Moeckel et al. (2018).

Twenty-three applications extracted from these four articles and the editorial book were then identified for further investigation. This led us to pinpoint the following five model requirements:

a   Has the potential of being applied to Singapore.
b   Ease of setting up and running with intuitive model elements.
c   Supported by a strong and active research team.
d   Fully accessible to allow for adaptation, extension, and incorporation of findings from other workstreams in the NUK programme.
e   Potential of being integrated with other models developed by other third parties.

Out of the 23 applications reviewed, only 3 were deemed suitable for the NUK programme. The following three applications were chosen for further investigation, based on the following criteria:

• SILO (http://silo.zone): Open-source and relatively simple to implement and developed to be integrated with MATSim (https://matsim.org) for an integrated land-use-transport-interaction model. Extensive work has been conducted in developing a MATSim model for Singapore, including gathering and processing the input data, in ways similar to those required for an RLCM.

- UrbanSim (http://www.urbansim.com): Open-source and the most widely used and well-established integrated land-use-transport-interaction model.
- SimMobility (https://mfc.mit.edu/simmobility): Integrated land-use-transport-interaction platform currently being developed for Singapore by the Future Urban Mobility, by the Singapore MIT Alliance for Research and Technology. The platform was being developed for evaluating future transport scenarios.

Publicly available information and the opinions of leading experts have been used to conduct the SWOT analysis presented in the following section.

### SILO: SWOT Analysis

SILO (Moeckel et al., 2018) is still relatively new and has been developed by Rolf Moeckel, Technical University of Munich, one of the leaders in integrated land-use and transportation modelling. The available models have been implemented in Germany and South Africa. All models were developed in Java with its source code continuously updated and available from https://github.com/msmobility/silo/.

### UrbanSim: SWOT Analysis

UrbanSim (Kii et al., 2016; Waddell, 2010) was developed by Paul Waddell, Berkeley, and formally distributed through https://www.urbansim.com/. It has been applied in numerous cities throughout the world. The software has been developed in Python with its source code continuously updated and available from https://github.com/UDST/urbansim.

### SimMobility: SWOT Analysis

SimMobility (Adnan et al., 2016) was initiated by Moshe Ben-Akiva, Kakali Basak, Chris Zegras, and Joe Ferreira at MIT. It has been further developed specifically for Singapore by research teams at MIT and the Singapore MIT Alliance for Research and Technology. It is not known what programming language is used for the models, or if its source code will become publicly available.

A detailed SWOT analysis for SILO, UrbanSim, and SimMobility is shown in Table 7.1.

### Selection of Application

Among the three applications considered, SimMobility was disregarded due to its intricate nature and primary focus on transport analysis. This narrowed down the

**TABLE 7.1** SWOT analysis for SILO, UrbanSim, and SimMobility

| | Strengths | Weaknesses | Opportunities | Threats |
|---|---|---|---|---|
| SILO | a Relatively simple.<br>b Full access to source code to allow for adaptations and extensions.<br>c Supported by a strong and accessible research team.<br>d Open source can be extended and maintained, independently from current core developers.<br>e Developed in Java, a popular and well-established programming language.<br>f Direct access to existing model applications.<br>g Two leading experts Prof Rolf Moeckel and Prof Kai Nagel are actively involved in the project, including direct development. | a Research opportunity to update the models with unique HDB conditions.<br>b Can be linked with MATSim for an integrated-land-use-transportation interaction model for Singapore.<br>c Given the MATSim link, it is easier to directly use critical input data formatted by FCL for the Singapore MATSim model.<br>d Developers are looking for new applications, thus creating an opportunity to directly influence and steer its development. | a No formal usage documentation or tutorials.<br>b No commercial support, so research will have to be identified and appointed to maintain and expand the model after the NUK project.<br>c Still in a prototype phase, and yet to be applied to a city with validation. | a Will initially be highly dependent on informal support from lead developers in Munich.<br>b Only two active applications by the core development team, without which development will stall. |

*(Continued)*

**TABLE 7.1** (Continued)

| | Strengths | Weaknesses | Opportunities | Threats |
|---|---|---|---|---|
| UrbanSim | a Well-established and used by multiple research groups in different regions over the world. <br> b Continuously being updated and improved by different research groups. <br> c Full access to source code to allow for adaptations and extensions. <br> d Open source, so can be extended and maintained, independently from current core developers. <br> e Developed in Python, currently the most popular programming language. <br> f Direct access to existing model applications. <br> g Formal commercial support is available. <br> h Documentation and tutorials available. | a Complex to set up and run, though effort has been invested by the core developers to improve this. <br> b Big push to using formal commercial services when dealing with actual software applications. <br> c Input data has to be formatted for the software from scratch. <br> d Other modules for the fully integrated land-use-transportation model have to be implemented from scratch. | a Will represent the first application of UrbanSim to Singapore. | a Investments will be required for commercial support, after which HDB will be locked in with the model. |

*(Continued)*

**TABLE 7.1** (Continued)

| | Strengths | Weaknesses | Opportunities | Threats |
|---|---|---|---|---|
| SimMobility | a Supported by strong MIT and Singapore research groups. <br> b Developed specifically for Singapore. | a Very complex and resource intensive to set up. <br> b Developed for transport analysis, not for residential location choice analysis. <br> c Residential location choice modules are not easily adaptable for the NUK project. | a Extension of RLCMs for HDB applications. | a Maintained by MIT and Singapore-MIT Alliance for Research and Technology (SMART) groups, without whom the project will fail. <br> b Source code not directly accessible and will be a black-box if used. <br> c Developed for transport analysis thus limits its application to NUK projects. |

*Source*: Authors.

options to UrbanSim and SILO for potential implementation. UrbanSim, while extensively documented, lacked connections to the Singapore context, thereby necessitating a complete setup from the ground up. On the other hand, SILO, despite being less established and in the developmental phase, offered the advantage of integration with existing models for Singapore. Furthermore, SILO had the capability to directly import formatted data from the Singapore MATSim model developed by the ETH Future Cities Lab in Singapore, thereby making it a viable and convenient choice for our project.

As a final test, both applications were downloaded and tested using their documented example cases. Whereas an UrbanSim model could be set up and run in under 15 minutes, setting up a SILO model was extremely challenging, and was abandoned after one day's worth of effort. Due to its newness, the SILO project is continuously being updated by its development team. Its example models are no longer maintained, and its documentation is also outdated. Given UrbanSim's better current strengths, and ease to set up, it was chosen as the preferred application to develop the ABRLM from. The only challenge with this selection was that it does not have a built-in population dynamics module, but it can be linked with any existing module. The model developed by Geard et al. (2013), Sim-Demog, was used and expanded for this purpose. Similar to UrbanSim, the model is freely available from https://github.com/nicgeard/sim-demog and accessible for extension. It was also developed in the same programming language, Python, as UrbanSim.

### Conceptual Design: Population Dynamics and Residential Relocation Model

The three main components for the ABRLM are:

a  An initial synthetic population.
b  A population dynamics model.
c  A residential relocation model.

In this section, a detailed discussion of the three core components ensues, accompanied by conceptual designs rooted in the model logic of these components. The section provides an in-depth overview of the basic designs, input data, and any necessary extensions for each component. A summary of these conceptual designs and their interconnections is illustrated in Figure 7.3. Prior to delving into the detailed designs, the section addresses two crucial project limitations and assumptions.

#### Model Limitations and Assumptions

For the NUK programme, the virtual population has been developed only for the following three designated towns in Singapore:

| MODELS | DESCRIPTION | DEVELOPMENT |
|---|---|---|

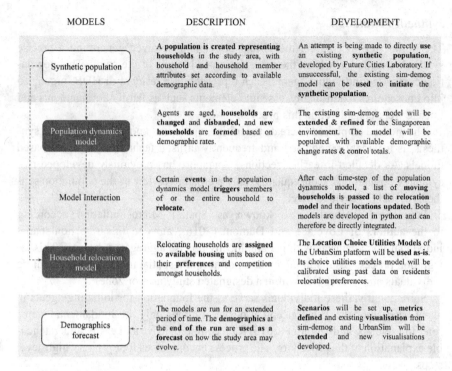

**FIGURE 7.3**   Virtual population conceptual design.

*Source*: Authors.

a   Toa Payoh.
b   Jurong East.
c   Punggol.

This inevitably led to a decrease in accuracy compared to a comprehensive model of Singapore. Most relocations occurred between areas outside these three towns and within them, with minimal movement between the towns. To reach its full potential, an alternative approach could have involved modelling the three towns as a closed system, enabling the utilisation of the complete relocation model but with unrealistic parameters. Despite these limitations, this method allowed for critical validation of the model logic.

Data from the other workstreams of the NUK programme, along with information provided by HDB and publicly available data, were employed to develop and validate the model. In cases where specific data was unavailable, assumptions were made, and international studies provided supplementary data.

Finally, a sensitivity analysis was performed to assess the impact of the absent input data on the ultimate outcomes of the model.

## Synthetic Population

The initial step in creating the virtual population involved generating a precise synthetic population within the three designated towns of the study. This population comprised individual agents, each representing a resident of HDB, organised into households. Additionally, essential elements such as family arrangements and lineage were considered in this process.

The method employed to create the initial population, known as population synthesis, is a widely recognised and frequently utilised technique for agent-based models. As detailed in earlier sections, Singapore has previously developed two synthetic populations. Consequently, only a broad overview of the population synthesis model is provided here.

Population synthesis, also known as Spatial Microsimulation according to the authors of Lovelace and Dumont (2016), aims to integrate non-spatial individual-level datasets with geographically precise aggregated datasets. This integration process results in the creation of accurate spatial microdata representing individuals and households within a designated study area or zone.

Subsequently, these individuals serve as the foundation for initiating agents in agent-based models. The conceptual design outlines only the fundamental steps; refer to Lovelace and Dumont (2016) for a comprehensive and easily understandable explanation of the procedure. The process encompasses four primary high-level steps:

*Step 1: Data Preparation.* The input for this process comprises a representative micro-sample of individuals and households, encompassing essential attributes required for the model, such as age, gender, ethnicity, annual income (or total annual household income), education level, and relationships with other members of the micro-sample if organised by households. The second input set consists of aggregate-level data known as control totals, indicating the total number of individuals within each town with specific attribute levels. For instance, these control totals might specify the total number of female residents within specific income brackets. These two input datasets are then converted into compatible formats tailored to the programming language employed for the population synthesis process.

*Step 2: Iterative Proportional Fitting:* After preparing the two input sources, a weighting algorithm is applied to determine how well each person in the micro-sample represents the corresponding geographical area with available control totals. The Iterative Proportional Fitting procedure (IPF) is a commonly used method in this phase. This procedure computes and assigns weights to combinations of individuals to match the control totals from census data. The likelihood of an individual belonging to a specific zone is then calculated and utilised to allocate individuals to respective zones. For instance, if Individual A has a calculated likelihood of 0.1 of being in Zone 1, and Zone 1 has 1,000 residents, 100 individuals with the same attributes as Individual A will be assigned to that zone.

*Step 3: Household Allocation.* As Step 2 solely assigns individuals to zones, an additional step is necessary to assemble these individuals into households. Several methods are available for this purpose. In a straightforward scenario where control totals for households are accessible, individuals are randomly selected and matched to create households in accordance with the specified control totals. Alternatively, if the micro-sample includes household details, individuals with their complete characteristics can be generated and organised using combinatorial optimisation techniques. This process ensures that each person is linked to one and only one household, minimising discrepancies between the synthetic population aggregates and the provided control totals.

*Step 4: Allocation of Households to Properties.* In the ultimate phase, a residential location model comes into play to allocate households to specific dwelling units, akin to the relocation model applied later in the virtual population model. This model functions as a discrete choice model, forecasting, through probability assignments, which unit each household is likely to select as their residence based on their unique socio-demographic characteristics and the attributes of the available units. The input for this model typically comprises disaggregate choice data, acquired through a combination of revealed preference and stated preference surveys. Alternatively, micro-level spatial data showing how residents moved between units over time can be employed to set up the model, provided that critical unit and household attributes are encompassed within the dataset. Following the model's establishment, households are allocated according to their assigned probabilities. This allocation can be carried out iteratively, commencing with the household possessing the highest probability, or randomly based on the allocation probabilities.

The input data required for the population synthesis, with its custodians were:

a  Aggregate population data for each town: publicly available from the Department of Statistics, Singapore.
b  Representative micro-sample: available from the LTA Household Travel Survey.
c  HDB flat attributes: publicly available from the Department of Statistics, Singapore.
d  Revealed and stated location preferences: obtained via survey data from the NUK programme.

## Population Dynamics Model

The objective of the population dynamics model is to replicate the fundamental life cycle of agents and households over a specified time span. For this purpose, an existing model, Sim-Demog, has been expanded and implemented. Every agent, generated during the population synthesis input phase, undergoes simulation at yearly intervals and is subject to key life events probabilistically. These events are modelled as Markov-chain processes. The underlying model logic, sourced from Geard et al. (2013), is outlined as follows:

1  Create an initial population, according to initial age distribution and household size distribution.
2  Scale all probabilities and rates to give rates per time step (year).
3  At each time step, each individual has the possibility of experiencing a life event as follows (individual attributes that affect event probabilities are listed in parentheses):

   •  Test for death (age, sex). If death occurs, the following occurs:

      i  The dead individual is removed from the population. If this results in a household containing orphaned children, any children who are old enough (e.g., >18 years) leave home, as per below, while any younger children are randomly allocated to another family household containing at least one child;
      ii If the option is activated, the birth of a replacement individual is triggered, and a mother is chosen (age, sex, parity, time since last birth).

   •  Test for couple formation (age, sex). If couple formation is to occur, select a partner from the population with an appropriate age difference and update their households as follows:

      i  If both individuals live at home, they move into a newly created household;
      ii If either (or both) of the individuals have their own household, the other partner (together with any dependents) joins them in this household.

   •  Test for leaving home (age). If an individual leaves home, they form a single-person household.

   •  Test for couple separation (age). If a couple separates, one individual remains in their former household, together with any dependents, while the other individual leaves to form a new single-person household.
   •  Calculate the number of additional births due to natural increases. Mothers are chosen for each of these new individuals as above.
   •  Calculate the number of new arrivals due to immigration. Immigrants arrive as a household unit, with the size of the household and the age of its occupants drawn from the current population age and household size distributions.
   •  Repeat Step 3.

If an accurate initial population, obtained through population synthesis, is accessible, the model operates using the existing population established in Step 3. However, if such data is not available, the model reverts to Step 1, potentially incorporating new births to replace deceased individuals. In this state, the model operates over an extended duration to create the initial population with authentic household structures. Subsequently, the model is run without the option of death replacement, potentially including immigration, for the designated number of simulation years.

The input data required for the population dynamics model, with its custodians are:

a   Household size and age distributions: publicly available from the Department of Statistics, Singapore.
b   Age- and sex-specific death rates: publicly available from the Department of Statistics, Singapore.
c   Age- and relationship status-specific birth rates: publicly available from the Department of Statistics, Singapore.
d   Age- and sex-specific marriage rates: publicly available from the Department of Statistics, Singapore.
e   Distribution of groom-bride age differences: publicly available from the Department of Statistics, Singapore.
f   Age- and sex-specific divorce rates: publicly available from the Department of Statistics, Singapore.
g   Probability of single person leaving home: unknown availability.

Several model extensions are necessary to accomplish this. Specifically, individual attributes, including race and income, need to be incorporated, and the model should accommodate the formation of multi-generational households. To account for these additional attributes, conditional probabilities must be obtained, especially if they differ from general probabilities. For instance, while the developed model relied on age- and sex-based death rates, more detailed and attribute-specific probabilities could be required if mortality rates vary significantly based on income levels and race. Estimations for the probabilities of forming multi-generational households also need to be made. If these values are not readily accessible, data from relevant countries could be utilised, and the model's overall sensitivity to these specific input parameters would then be rigorously tested.

### Household Relocation Model

When the population dynamics model triggers a change in household composition, the newly created households are marked for potential relocations. New HDB households coming into the study area are also marked for relocation. Available properties in the area are then updated, and the households are assigned to the properties based on availability, property attributes, and household attributes.

For this model, the existing Statistics, Supply and Demand Accounting, Transition and Relocation and Location Choice Utilities models within the existing UrbanSim platform have been used.

Figure 7.4, taken from Waddell (2010), shows a high-level description of the residential location model of UrbanSim. For the virtual population, the model starts

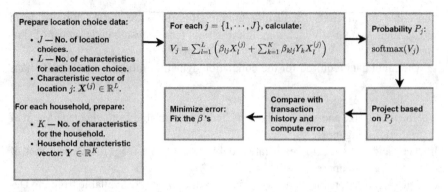

**FIGURE 7.4** UrbanSim RLCM logic (from Waddell, 2010).

*Source*: http://udst.github.io/urbansim/models/index.html, last accessed 23 April 2019.

after the Get choosers step shown in the bottom of the figure. The choosers are passed as input from the population dynamics model.

First, households that will relocate are linked with available housing units through the Choice Set Generation step. Next, each household's characteristics, such as income and household size, as well as characteristics of the linked available housing units, such as density, age, and accessibility to employment and other opportunities, are combined via a Multinomial Logit Utility Equation to calculate the attractiveness (utility) of each unit per household. The utility values are then converted into selection probabilities using the Multinomial Logit Probability Equation. The model then randomly simulates, based on the calculated probability, the households choosing a specific unit. If multiple households choose the same unit, different mechanisms can be used to allocate a final household, such as first assigned, random assignment, or selecting the household with the highest income. Through this process, households are allocated to available units.

The input data required for the relocation model, with its custodians are:

a   Revealed location and unit preferences, used to define the utility scores of units to households: obtained via the NUK survey and HDB data on the actual relocation of HDB residents.
b   Stated location and unit preferences, used to define the utility scores of units to households: obtained via the NUK survey.

Extensions required for the module includes adding constraints for racial quotas per town and block level, resell restrictions based on HDB regulations, incentive schemes to make the housing market accessible to key demographic groups, such as young first-time parents, and to incentivise households to relocate to more convenient locations, for example, to units closer to immediate family members.

The relocation model can consider the populations and the available residences for relocation only within the towns represented within the simulation.

To make this model more successful it is advisable to extend the concept to all of Singapore.

The steps that are involved in simulation that tracks demographic changes and residence relocations can be summarised as follows:

Step 1.  Load the data required for demography simulation.
Step 2.  Load the historical data of households that had relocated in the past and the residences that they had chosen together with all attributes of the households and the residences.
Step 3.  Determine the $\beta$ coefficients in the multinomial logit relations shown in Fig. 7.4 from the historical data by error-minimisation method.
Step 4.  Choose $\Delta T_1$ as the time step for demography simulation.
Step 5.  Choose $\Delta T_2 = n \Delta T_1$ as the time step for relocation simulation, where $n$ is a chosen positive integer.
Step 6.  Evolve the population by the time step $\Delta T_1$ for $n$ times by the demography simulation, Sim-Demog.
Step 7.  Find the households that need relocation of the residences, and form a *feature table* for these households (a *feature table* is a table that contains the attributes as the columns).
Step 8.  Find the residences that are available for occupying and form a feature table of these residential units.
Step 9.  Carry out the relocations using the multinomial logit scheme in Figure 7.4.

Iterate Steps 6–9 for the whole duration of the simulation.

## Specifications of Virtual Population ABM Model

In our modelling and simulations, we employed the Sim-Demog model crafted by Geard and his team. Within this framework, every individual is characterised by attributes like age, gender, marital status, type of family, employment status, home ownership, and number of children. These attributes undergo dynamic shifts over time, with some alterations being steered stochastically based on event probabilities such as birth, death, marriage, and divorce, all reflective of Singapore's context. Additionally, the model maintains a record of interpersonal relationships, like the connection between husband and wife, or father and son, among others.

In the original model, the starting population is typically generated by a simulation that aligns demographic characteristics, such as age distribution, with census data observations. However, our approach deviates by drawing the initial population from an extensive survey conducted in three distinct locations: Jurong East, Punggol, and Toa Payoh. These towns comprise multiple neighbourhoods, with each town having between six and eight neighbourhoods. The survey encompassed a total of 2,580 households. While the survey had only one respondent per household, the simulation considered every household member as disclosed by that respondent.

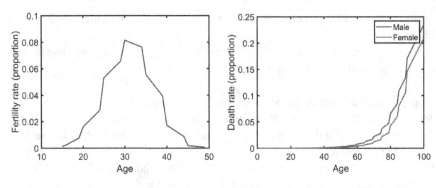

**FIGURE 7.5** Death rates and fertility rates versus age. Top: The proportion of individuals dying between an age $x$ and $x + 1$. Bottom: The proportion of women giving birth in an age between $x$ and $x + 1$ (obtained from, 'Population Trends,' Department of Statistics, Singapore, 2020).

*Source*: Authors.

Using an annual time step, events like birth, death, marriage, divorce, and the establishment of new households when an individual reaches adulthood and departs from their parents' home are executed based on probabilities derived from census data provided by the Department of Statistics, Singapore. The age-specific mortality and fertility rates applied in the simulation can be seen in Figure 7.5. The model also recognises the four main ethnic groups in Singapore: Chinese, Malay, Indian, and others. It has been adjusted to factor in the likelihood of inter-ethnic marriages during its temporal progression.

The yearly likelihoods for events like marriage and divorce represent the chances of an individual getting married or divorcing within a year, respectively. In the simulation, an adult has the option to establish a one-person household without getting married. The annual likelihood of such an occurrence in the context of Singapore can be specified for marriage, divorce, and for the event of an adult leaving home, and can be given as the probabilities 3.74%, 0.74%, and 0.8%, respectively. These figures have been derived from the publications of the Department of Statistics, Singapore, for the year 2020. Additionally, 5% of births are attributed to single mothers, as indicated. Inter-ethnic marriages make up 17.6% of the total marriages. These odds are integrated into the simulation through random number generators at every annual step. When inter-ethnic marriages occur, the ethnic backgrounds of the partners are determined based on the probabilities found in the census data from the Department of Statistics, Singapore.

It's important to highlight that the yearly probabilities of marriage and divorce might vary based on the partners' ages, indicating potential enhancements for this model. In our simulation, there are age limits set for marriage and divorce. Partners must be at least 18 years old and no older than 60 years of age to experience marriage or divorce events. Moreover, age-specific rates are applied for mortality and fertility. For mortality rates, distinctions between genders are made as per the 2020 census data. The simulation also factors in the age gap between male and female partners during couple formation, viewing this difference as a normal distribution with a specified mean and standard deviation.

## Results at the Town Level

The simulations were carried out with the initial population obtained from the large-scale NUK survey. The demographic changes for a period of 20 years are presented in this section for the three HDB towns, namely, Jurong East, Punggol, and Toa Payoh. Though these simulations were performed at the scale of neighbourhood, we have aggregated the results to the town level. Figure 7.6 shows the following for Jurong East: the distribution of the fraction of the individuals with respect to their age (panel a), the distribution of the fraction of the households with respect to the household size in (panel b), the age composition with respect to the household size in (panel c), the age composition with respect to the household age in (panel d), the variation of proportions of the family types over years (panel e), the variation of the proportions of the household sizes over years (panel f), and the proportions of individuals of different age groups in each of the family types. The panels (a) and (b) show the respective data at the beginning and the end of the 20 years simulation. The panels (c) through (f) show the snapshot of the population at the end of the 20 years' simulation.

As can be seen from Figure 7.6a, the overall population gets older as years evolve. Figure 7.6b describes that the proportions of the single-member households and that of the couples not living with children increase over the years, and that the proportions of the households of sizes three and above fall over the years. These trends in panels (a) and (b) highlight the fact that the total fertility rate is below the required threshold level for replacement of the population.

The age composition with respect to the household size and the household age is given in Figure 7.6c and d. The household age is defined as that start from the time of beginning of the simulation, or from the time of constitution of a newer household due to marriage between two individuals, or due to individuals leaving their previous households at certain probability to form new households when they become adults. Figure 7.6c shows that the single-member households are mostly older individuals of age above 65, and that the young children or young adults are in the households which contain at least three members. Figure 7.6d shows that higher the household age, more likely they will contain individuals above 65. This phenomenon is caused by the formation of new households by members upon marriage or due to the event of leaving home by individuals that form single-member households.

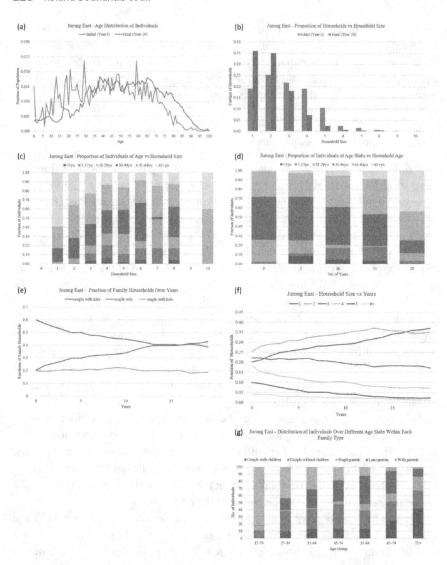

**FIGURE 7.6** Individual age, household, and family type changes for Jurong East. (a) Age distribution of individuals, (b) proportion of the households versus household size, (c) proportion of individuals of age versus household size, (d) proportion of individuals of age slabs versus household age, (e) fraction of family households over years, (f) household size versus years, (g) distribution of individuals over different age slabs within each family type.

*Source*: Authors.

The time evolution of the proportions of different types of families of sizes greater than one are shown in Figure 7.6e. It shows that the proportion of the family that contains both partners and children decreases over years, while the households that contain only partners without children living with them increase. The households that contain a single parent and his/her child/children remain without undergoing significant change in the proportion over the years.

The temporal variation of the proportions of the households of different sizes is shown in Figure 7.6f. It shows that the households of sizes greater than two drop in their proportions over the years, while the single-member households and couples without children living in them keep increasing in their proportions during the same period. This phenomenon agrees with the observations made in Figure 7.6b. This is, as explained above, due to the low total fertility rate in Singapore, and since the children as they grow up tend to form a new household either by marriage or by forming a single-member household at certain probability for this event upon reaching an adult age.

In Figure 7.6g, the couple-with-children family type has a peak in the age group of 45–55. This can be explained as follows. At this middle age, the individuals have a higher chance of having children born in any previous years. Beyond this age of the individuals, their children start becoming adults and form new households either through moving out as a single person or through marriage.

Furthermore, the younger the individuals, the more likely they are to live with their parents as shown in the last sub-panel of Figure 7.6g. Similarly, the older the individuals, the more likely they are to constitute a single-member household as shown. Further, the older the individuals, the more likely they are to constitute a single-parent household, as shown in Figure 7.6g, since there is a higher chance for the death of one of the partners, or for a divorce between them.

Figures 7.7 and 7.8 show the results for Punggol and Toa Payoh, but as in Figure 7.6. In Figure 7.7a, the age distribution can capture the double peaks that we saw in Figure 7.1, which is the characteristic of a town that contains young couples such as Punggol. The time evolution is found to preserve this distribution approximately but shifts to greater ages. The rest of the results in Figures 7.7 and 7.8 exhibit good similarity with Figure 7.6.

In Figure 7.9, we show the temporal evolutions of the eight resident archetypes identified in Chapter 2. These eight clusters were found by the method of k-medoid in conjunction with the large-scale survey data among the participants. The statistical information about these clusters was obtained from the earlier workstream and used for mapping every individual in the simulation to those clusters in a many-to-one manner. Figure 7.9a–c shows these for Jurong East, Punggol, and Toa Payoh, respectively.

In all the three towns, cluster 1, which is characterised by couples-only family type with a mean age of 71, experiences an overall growth with respect to years.

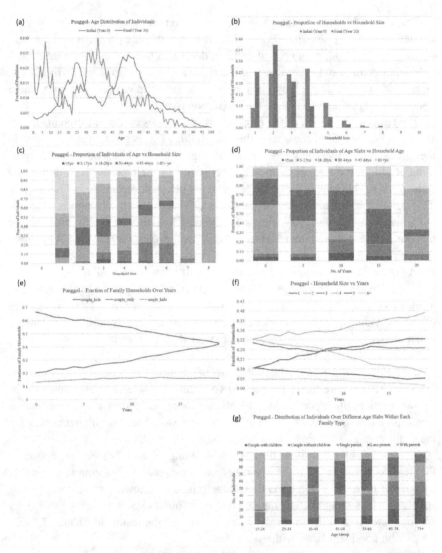

**FIGURE 7.7** Individual age, household, and family type changes for Punggol. (a) Age distribution of individuals, (b) proportion of the households versus household size, (c) proportion of individuals of age versus household size, (d) proportion of individuals of age slabs versus household age, (e) fraction of family households over years, (f) household size versus years, (g) distribution of individuals over different age slabs within each family type.

*Source*: Authors.

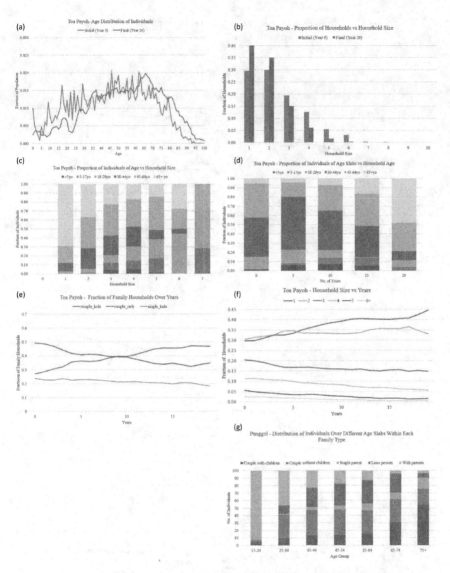

**FIGURE 7.8**   Individual age, household, and family type changes for Toa Payoh. (a) Age distribution of individuals, (b) proportion of the households versus household size, (c) proportion of individuals of age versus household size, (d) proportion of individuals of age slabs versus household age, (e) fraction of family households over years, (f) household size versus years, (g) distribution of individuals over different age slabs within each family type.

*Source*: Authors.

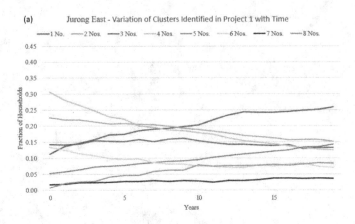

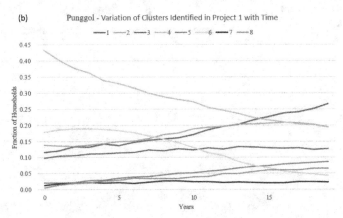

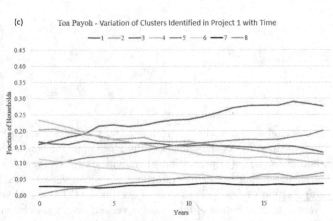

**FIGURE 7.9** Variation of clusters identified in Project 1 with time. (a) Jurong East, (b) Punggol, (c) Toa Payoh.

*Source*: Authors.

It is vice versa in the case of cluster 4, which represents young couples with children. Since the number of families of this type is relatively higher in Punggol, it undergoes this decline at a higher rate than the other towns.

cluster 2 that is characterised by the family type of middle-aged couple with children underwent a growth in Punggol over the years while it declined in other towns. This can be understood from the fact that Punggol contains significant families of the young couple type with children forming cluster 4, and that these households would get aged and contribute to the proportions of cluster 2.

## Future Directions

The demography changes are not just due to temporal events such as birth, death, marriage, divorce, and so on, considered in this work. The changes due to spatial movements of the residence of the population significantly affect the demography at a chosen location. These spatial movements can be driven by various factors such as the proximity of the places of employment, children's schools, housing prices, individuals' income, and transport facilities. The attributes of the houses such as the number of bedrooms, baths, and sizes of the houses are also parameters in deciding the chance of relocation of a household to a new house. In Singapore's context, the ethnicity quota on public housing needs to be considered to model the ethnic makeup of a neighbourhood more accurately.

For this purpose, we intend to consider the relocation model as described by Waddell as our future direction of study. Under this scheme of Waddell, at every time step of the demographic simulation, the households that require relocations caused by the above-mentioned factors will need to be identified. Further, several alternative options of houses available for purchase by these households will also need to be identified as part of this step of the simulation. The probability of a household relocating to each of the available options is given by a discrete choice model, such as multinomial logits. However, the scaling factors that determine the impact of each of the parameters in the above paragraph need to be obtained from a statistical fitting on the historical data of past purchase transactions using multinomial logits as implemented in the Python package UrbanSim developed by a group headed by Waddell. Then the relocations are carried out based on these probabilities.

## Conclusion

The demographic simulations based on the input from the large-scale survey conducted in three HDB towns, namely, Jurong East, Punggol, and Toa Payoh, were performed by making use of 2020 census data for mortality and fertility rates prevailing in Singapore.

The demographic changes predicted by agent-based simulation for the three types of towns in Singapore indicate various facts: the proportions of single- and two-member households increase monotonously with respect to time across all three types of towns. The proportions of the households of size greater than two drop over the years. The distribution of the proportions of individuals with respect to age highlights the issue of low total fertility rate. These facts are also reflected in the time evolution of the family types and household sizes and the variation of age compositions with respect to household sizes and household ages. Across all three types of towns, the proportions of individuals belonging to the family type of couples that are not living with children increase over the years, and the proportions of individuals belonging to the family type that consists of couples only decrease over the same years.

The results of this model can be utilised to study different scenarios by changing the input parameters at our disposal. Such scenario studies reflect on the household compositions under the time evolution, which could help, for example, get a grasp of family types that most Singaporeans of certain age groups would belong to.

This virtual population model can be further enhanced by considering age-specific probabilities for the events marriage and divorce. Currently this is treated as a constant for all ages between a lower and an upper cut-off value for the ages. This model also needs to include immigration and emigration events.

One should note that this model has been developed under the framework of Python's object-oriented programming. The main object of the simulation is 'Population' that contains the objects such as households and individuals. The residences-related data can only go as attributes of the individuals or households in the population. This results in some complexity in defining spatial information such as towns and neighbourhoods in the simulation. This issue can be addressed by overhauling the data structure of the simulation. Changes are needed to re-structure the objects such that the hierarchy is as follows: Country -> Districts -> Towns -> neighbourhoods -> residential buildings -> units -> households -> individuals. Such a redefinition will require a considerable amount of time, but this should be the direction to make this model integrate well with a residential relocation model.

The more accurate demography change will need to consider the spatial dynamics of the residences of the individuals through an appropriate resident relocation model. This, however, requires a vast amount of data on the individuals and the real estate, including the past transaction history.

### Possible Use in Planning Practices

Virtual population models can aid the urban planning process in significant ways. By integrating these models into the planning process, cities can better understand the complex interactions between population dynamics, land use, and infrastructure, leading to more holistic, informed, and effective decision-making in planning.

Virtual population models can simulate and predict the growth, migration patterns, and demographic shifts within urban areas. By analysing historical data and trends, planners can use these models to predict future changes in population size, age distribution, household composition, and other key factors. For example, a city planner might use a virtual population model to forecast the influx of young professionals into a neighbourhood, allowing them to anticipate the need for different amenities and services. Similarly, a city facing an ageing population might use a virtual population model to assess the demand for senior housing, healthcare facilities, accessible transportation options, and other strategies.

These models can also help identify tipping points or thresholds at which certain demographic changes may occur. By simulating different scenarios and analysing the effects of various factors (such as housing prices, employment opportunities, or public policy changes), planners can pinpoint critical thresholds that could lead to significant shifts in population dynamics. For instance, a virtual population model might reveal the point at which rising housing costs trigger an exodus of lower-income residents from a neighbourhood. In such situations, by identifying these disparities early on, planners can implement policies and initiatives to prevent demographic challenges, such as gentrification, displacement, or social exclusion.

An important aspect to consider when utilising virtual population models is translating the learnings from such models spatially to the allocation of amenities and other facilities during design and planning. In this process, virtual population models can help assess the spatial distribution of amenity demand within public housing estates based on population density, socio-economic indicators, and accessibility factors. Planners can use Geographic Information Systems (GIS) tools to overlay demographic data with existing amenities, identify underserved areas or population clusters, and visualise spatial patterns of demand. Spatial analysis techniques such as kernel density estimation, hotspot analysis, and spatial interpolation can help prioritise amenity allocation and determine optimal locations for new facilities. Thus, virtual population models can inform the optimal placement of amenities within public housing estates to maximise accessibility and usability for residents.

These models can also facilitate community engagement and stakeholder collaboration in the allocation of amenities within public housing estates. Planners can use interactive mapping tools, 3D visualisations, and virtual reality simulations to communicate amenity plans to residents, solicit feedback, and co-design solutions that meet community needs. Participatory planning processes can help build consensus, identify local preferences, and foster a sense of ownership and pride in the development of public housing estates.

By integrating virtual population models with spatial planning tools, data-driven analysis techniques, and participatory decision-making processes, planners can effectively allocate amenities during the design and planning of estates to create vibrant, equitable, and inclusive communities.

To make all these possibilities a reality, the generalizability of such models across different geographic regions and socio-cultural contexts needs to be ensured. A highly important consideration is the availability of high-quality data. Virtual population models rely on data sources such as census data, surveys, satellite imagery, and administrative records. For reliable applications, it's essential to have access to reliable and comprehensive data sources that capture key demographic, economic, and spatial variables. Data quality can vary significantly between countries and regions, so efforts may be needed to improve data collection methods and ensure consistency and reliability of the results.

Additionally, demographic behaviours and patterns can widely vary between places due to cultural differences and socio-economic and historical factors. When adapting virtual population models for different regions, it's crucial to consider these contextual differences and tailor the model parameters, assumptions, and application contexts accordingly. For example, household structure, fertility rates, migration patterns, and transportation preferences may differ between cities and countries, requiring a context-specific calibration and validation of the model.

The regulatory and policy landscape can also vary significantly between countries, affecting factors such as land-use planning, housing affordability, transportation infrastructure, and social services provision. When adapting virtual population models for different applications, it's important to consider the local policy and governance context and incorporate relevant regulations, incentives, and constraints into the model framework. This may involve collaborating with local stakeholders, government agencies, and policymakers to ensure the model's relevance and usefulness for decision-making.

Finally, before applying virtual population models in different contexts, it's essential to validate the model's accuracy and reliability using empirical data and independent validation techniques, such as sensitivity analysis.

## Acknowledgement

The virtual population model developed in this research has been integrated into the City Application Visualisation Interface (CAVI) platform developed by Électricité de France, EDF Lab Singapore. Through the CAVI platform, users can change the input parameters to view the projected trends in each town, building on this virtual population model. The authors are grateful to EDF Lab Singapore for supporting this research.

## Bibliography

Acheampong, R. A., & Silva, E. A. (2015). Land use–transport interaction modelling: A review of the literature and future research directions. *Journal of Transport and Land Use, 8*, 11–38.

Adnan, M., et al. (2016). Simmobility: A multi-scale integrated agent-based simulation platform. [Unpublished manuscript].

Arentze, T., & Timmermans, H. (2010). The residential choice module in the Albatross and Ramblas model systems. In F. Pagliara, J. Preston, et al. (Eds.), *Large-scale agent-based transport demand model for Singapore*. [Unpublished manuscript].

Beykaei, S., & Miller, E. J. (2017). Testing uncertainty in ILUTE–An integrated land use-transportation micro-simulation model of demographic updating. *Journal of Civil Environmental Engineering, 7*, 1–9.

Brown, D. G., & Robinson, D. T. (2006). Effects of heterogeneity in residential preferences on an agent-based model of urban sprawl. *Ecology and Society, 11*(1), 46. http://www. ecologyandsociety.org/vol11/iss1/art46/

Buchmann, C. M., Grossmann, K., & Schwarz, N. (2016). How agent heterogeneity, model structure and input data determine the performance of an empirical ABM–a real-world case study on residential mobility. *Environmental Modelling & Software, 75*, 77–93.

Department of Statistics, Singapore. (2017). *Population Trends 2017*. Singapore: Ministry of Trade & Industry, Republic of Singapore.

Department of Statistics, Singapore. (2020). *Population Trends, 2020*. https://www.singstat. gov.sg/-/media/files/publications/population/population2020.pdf

FCL. (n.d.). *Engaging mobility*.

Feldman, O., et al. (2010). A microsimulation model of household location. In F. Pagliara, J. Preston, & D. Simmonds (Eds.), *Residential location choice: Models and applications* (pp. 223–242). Berlin, Heidelberg: Springer.

Fourie, P. (n.d.). *Future cities laboratory*.

Geard, N., McCaw, J. M., Dorin, A., Korb, K. B., & McVernon, J. (2013). Synthetic population dynamics: A model of household demography. *Journal of Artificial Societies and Social Simulation, 16*(1), 8.

Gregor, B. (2007). Land use scenario developer: Practical land use model using a stochastic microsimulation framework. *Transportation Research Record: Journal of the Transportation Research Board, 2003*, 93–102.

Heldt, B., Gade, K., & Heinrichs, D. (2016). Determination of attributes reflecting household preferences in location choice models. *Transportation Research Procedia, 19*, 119–134.

Huang, Q., Paelinck, J., & Shen, S. (2014). A review of urban residential choice models using agent-based modelling. *Environment and Planning B: Planning and Design, 41*(4), 661–689.

Huang, Q., Parker, D. C., Sun, S., & Filatova, T. (2013). Effects of agent heterogeneity in the presence of a land-market: A systematic test in an agent-based laboratory. *Computers, Environment and Urban Systems, 41*, 188–203.

Hunt, J. D., & Abraham, J. E. (2010). Household behaviour in the Oregon2 model. In F. Pagliara, J. Preston, & D. Simmonds (Eds.), *Residential location choice: Models and applications* (pp. 181–208). Berlin, Heidelberg: Springer.

Hunt, J. D., Kriger, D. S., & Miller, E. J. (2005). Current operational urban land-use--transport modelling frameworks: A review. *Transport Reviews, 25*, 329–376.

Joubert, J. W. (n.d.). *Industrial and systems engineering*. Retrieved from https://www.up.ac. za/en/industrial-and-systems-engineering/article/2439902/prof-johan-w-joubert

Kii, M., Nakanishi, H., Nakamura, K., & Doi, K. (2016). Transportation and spatial development: An overview and a future direction. *Transport Policy, 49*, 148–158.

Lovelace, R., & Dumont, M. (2016). Spatial microsimulation with R (1st ed.). Chapman and Hall/CRC. https://doi.org/10.1201/9781315381640

Macal, C. M., & North, M. J. (2010). Tutorial on agent-based modelling and simulation. *Journal of Simulation, 4*, 151–162.

Manson, N. J. (2006). Is operations research really research? *Orion, 22*, 155–180.

MIT. (n.d.). *Software*. Retrieved September 23, 2018, from https://mfc.mit.edu/simmobility

Moeckel, R. (2017). Constraints in household relocation: Modelling land-use/transport interactions that respect time and monetary budgets. *Journal of Transport and Land Use, 10*, 211–228.

Moeckel, R., Garcia, C. L., Chou, A. T. M., & Okrah, M. B. (2018). Trends in integrated land use/transport modelling: An evaluation of the state of the art. *Journal of Transport and Land Use, 11*. https://doi.org/10.5198/jtlu.2018.1205

Müller, K., & Axhausen, K. W. (2010). Population synthesis for microsimulation: State of the art. *Arbeitsberichte Verkehrs-und Raumplanung, 638*, 1–15.

Munich School of Management. (n.d.). Retrieved from https://www.msm.bgu.tum.de/en/team/rm/

Munich School of Management. (n.d.). Retrieved September 23, 2018, from https://www.msm.bgu.tum.de/index.php?id=5&L=1

Pagliara, F., Preston, J., & Simmonds, D. (2010). *Residential location choice: Models and applications*. Berlin, Heidelberg: Springer.

Pidd, M. (2010). Why modelling and model use matter. *Journal of the Operational Research Society, 61*, 14–24.

Schelling, T. C. (1969). Models of segregation. *The American Economic Review, 59*(2), 488–493.

Schirmer, P. M., Van Eggermond, M. A. B., & Axhausen, K. W. (2014). The role of location in residential location choice models: A review of literature. *Journal of Transport and Land Use, 7*(2), 3–21.

Sun, L., & Erath, A. (2015). A Bayesian network approach for population synthesis. *Transportation Research Part C: Emerging Technologies, 61*, 49–62.

Sun, L., Erath, A., & Cai, M. (2018). A hierarchical mixture modelling framework for population synthesis. *Transportation Research Part B: Methodological, 114*, 199–212.

UrbanSim. (n.d.). Retrieved September 23, 2018, from https://www.urbansim.com

Van Eggermond, M. A. B. (2017). Diversity, accessibility and its impact on vehicle ownership and residential location choices. [Unpublished manuscript].

Van Eggermond, M. A. B., Erath, A., & Axhausen, K. W. (2018). Residential search and location choice in Singapore. In *2018 TRB Annual Meeting Online, 97th Annual Meeting of the Transportation Research Board (TRB 2018)*, Washington, DC, USA, January 7–11, 2018. https://doi.org/10.3929/ethz-b-000193146

Waddell, P. (2010). Modelling residential locations in UrbanSim. In F. Pagliara, J. Preston, & D. Simmonds (Eds.), *Residential location choice: Models and applications* (pp. 165–180). Berlin, Heidelberg: Springer.

Waddell, P. (2002). UrbanSim: Modelling urban development for land use, transportation, and environmental planning. *Journal of the American Planning Association, 68*(3), 297–314.

Wegener, M. (2011). From macro to micro-how much micro is too much? *Transport Reviews, 31*, 161–177.

Xu, Z., Glass, K., Lau, C. L., Geard, N., Graves, P., & Clements, A. (2017). A synthetic population for modelling the dynamics of infectious disease transmission in American Samoa. *Scientific Reports, 7*(1), 1–9.

Zhu, Y., & Ferreira Jr, J. (2014). Synthetic population generation at disaggregated spatial scales for land use and transportation microsimulation. *Transportation Research Record, 2429*, 168–177.

# PART III
# Enabling Inclusivity

Building on the deep understanding of diverse residential demography in Part I and the comprehensive urban analytics tools and processes in Part II, community-based participatory research was carried out in targeted intervention sites which led to the development of a Community Enablement Playbook.

DOI: 10.4324/9781003437659-11

# 8

# COMMUNITY ENABLEMENT PLAYBOOK

*Keng Hua Chong, Tshui Mum Ha, Denise Nicole Lim,
Natasha Yeo Min, Sze Min Neo, Sofia Foo Xin Di and
Sin Mei Cheng*

## Introduction

Since before Singapore's independence in 1965, community infrastructure such as Community Centres (CCs), Residents' Committees (RCs), Neighbourhood Committees (NCs), and more recently Residents' Networks (RNs) have formed the cornerstone of local community engagement and development in Singapore.[1] Initially conceived as communal spaces for social activities, especially for residents relocated from their original villages in the early years of independence, these community infrastructure evolved to encompass broader social programmatic roles, including fostering racial harmony, community cohesion, and emergency preparedness, as evidenced during the recent COVID-19 pandemic.

In contemporary Singapore, a myriad of factors, including demographic shifts, technological advancements, economic evolution, and democratic maturation, challenge the adequacy of traditional community infrastructure and programmes. Through the studies detailed in Chapters 1–3, it becomes apparent that Singapore requires a more adaptable and inclusive community support structure capable of addressing a diverse array of needs, ranging from eldercare and intergenerational programming to the integration of new immigrants and ensuring financial accessibility to community resources for all. The digital revolution has further revolutionised the landscape, offering novel opportunities for the integration of digital and physical spaces, thereby fostering new platforms for social interactions and community resilience.

As the city-state progresses, there is a burgeoning demand among its populace for more participatory forms of governance and community engagement, steering away from centralised approaches. Hence, there arises an urgent need for a novel framework of community enablement that is agile, inclusive, and adept at

DOI: 10.4324/9781003437659-12

leveraging technological advancements for the common good. Such a framework must not only address the current and emerging needs of Singapore's diverse population but also anticipate future challenges and opportunities. This chapter aims to present alternative approaches through which citizens can contribute to fostering vibrant and engaged neighbourhoods, where individuals and groups are empowered to contribute to community development.

This chapter embarks on an exhaustive exploration of the insights gleaned from existing place-making initiatives in Singapore, followed by the development and validation of various engagement strategies and toolkits across three distinct towns—Jurong East, Punggol, and Toa Payoh. These initiatives are grounded in a comprehensive understanding of resident archetypes, their perceived Quality of Life (QoL) and Quality of Place (QoP), and the diverse needs and assets of each neighbourhood, as elucidated in the preceding chapters. Leveraging the PAR approach,[2] the chapter delves into the opportunities and challenges encountered in appropriating spaces for community activities, while also delineating the role of community designers as mediators between local administrations and community-led endeavours.

Through the iterative process of prototyping and post-implementation reviews, the chapter dissects the multi-stakeholder dynamics and creative processes underlying ground-up appropriation and community design within the unique socio-political context of Singapore. The culmination of this endeavour is the 'Community Enablement Playbook,' a comprehensive guide designed to navigate the non-linear trajectory of community engagements and empowerment in diverse neighbourhood environments. Grounded in a data-driven co-creation process, the playbook aims to foster effective collaboration between residents, grassroots leaders, and local authorities, with the overarching goal of promoting creative place-making initiatives and nurturing enduring, self-sustained community-led place-keeping endeavours across Singapore.

## Case Studies

The research began with a comprehensive examination of 16 existing place-based community initiatives across Singapore (Figure 8.1 and Table 8.1), aiming to analyse the underlying multi-stakeholder relations and co-creative processes while identifying patterns of citizen-initiated appropriation and community design uniquely adapted to Singapore's socio-political context. Semi-structured interviews were conducted with active members of these initiatives to glean insights into key stakeholders involved, the initiative's developmental process, challenges encountered, resultant changes, sustainability factors, and its evolutionary trajectory, encompassing both growth and decline. Additionally, site observations were conducted to discern how common amenities and public spaces facilitated these initiatives. Each interview was meticulously transcribed and systematically analysed through coding, with the goal of identifying, categorising, and describing commonalities, differences, patterns, and structures found in the transcripts (Holton, 2007).

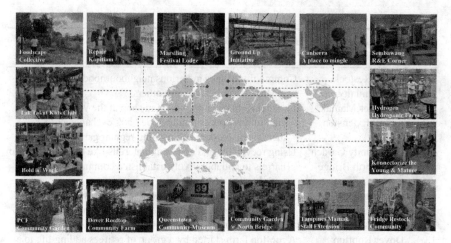

**FIGURE 8.1**   Locations of selected case studies of place-based community initiatives across Singapore.

*Source*: Authors.

**TABLE 8.1**   List of selected place-based community initiatives and descriptions

| No. | Selected Case Studies | Description of Community Initiatives |
|---|---|---|
| 1 | Foodscape Collective | Creating an equal, inclusive, and sustainable food system, through food-related practices such as farming, reducing wastage, etc. |
| 2 | Repair Kopitiam | Introducing repair efforts to the community, teaching them how to mend broken items, to tackle the buy-and-throw culture of today. |
| 3 | Marsiling Festival Lodge | Temporary decorative structure with lighting for different festivals built by a retired carpenter every year at the common space to promote inter-racial/religious harmony. |
| 4 | Ground-Up Initiative (GUI) | A modern-day village focusing on nature and community building through hands-on activities like farming, cooking, and wood crafts. |
| 5 | Canberra: A place to mingle | Revamping a communal area with seats and murals full of historical context and information, for children to learn about the neighbourhood. |
| 6 | Sembawang R&E Corner | Decorating an underutilised resident corner with art murals, transforming it into a conducive spot for gathering. |
| 7 | Hydrogen Hydroponic Farm | A neighbourhood hydroponic farm that promotes community bonding through farming activities. |
| 8 | Konnectorize the Young and Mature | Installing 3D wall murals titled 'From Farmland to a Modern City' by residents of various ages, creating a cosy hang-out for all. |

(*Continued*)

TABLE 8.1 (Continued)

| No. | Selected Case Studies | Description of Community Initiatives |
|---|---|---|
| 9 | Fridge Restock Community | Community fridges in various neighbourhoods for residents to share surplus food with others. |
| 10 | Tampines Mamak Stall Extension | Expanding an original provision shop at the void deck into community spaces that include book exchange corner, therapeutic garden, and running regular garden parties. |
| 11 | Community Garden @ North Bridge | Using recyclable and common materials like tyres and plastic barrels to collaboratively establish a community garden with local residents. |
| 12 | Queenstown Community Museum | A community museum that showcases the town's heritage initiated and run by volunteers with the support from the local community. |
| 13 | Dover Rooftop Community Farm | A rooftop farm started by a group of retirees adding life to an otherwise deserted rooftop and promoting neighbourliness in the area. |
| 14 | PCF Community Garden @ Jurong | A community garden for children from the PCF kindergarten to learn about nature, as well as a social space for residents. |
| 15 | Bold at Work | Empowering youth living in the vicinity with the mindsets and tools that they need at work and in their communities. |
| 16 | Tak Takut Kids Club | A children's club that aspires to provide a safe haven for children from vulnerable backgrounds. |

*Source*: Authors.

The in-depth analysis of each case study, coupled with comparative assessments, aims to offer clearer, more intimate perspectives of on-the-ground realities, thereby informing future real-life development of place-making and place-keeping initiatives. By uncovering the opportunities and challenges faced by Singapore grassroots leaders and community designers in appropriating spaces for community activities, the research aimed to uncover how community designers could assume the novel role of mediators to seamlessly bridge the gap between local administration and community-led creations.

Eight key strategies were found to be pivotal to the success of community initiatives in Singapore concerning place-making and place-keeping.

*Place-making*, an approach emphasising participatory design processes, communal vision translation, social interaction, and sustainable development, aims to create quality public spaces that contribute to people's health, happiness, and well-being (Project for Public Spaces website). The concept originated in the 1960s when urbanists like Jane Jacobs and William H. Whyte advocated for cities designed to prioritise people over vehicles and commerce. Four main strategies for successful place-making in Singapore were revealed through case studies (Chong, Gao & Kato, 2019):

1 *Triggers and discourse*: The motivation, viewpoints, and narratives of stakeholders.
2 *Actors*: The involvement of both internal and external stakeholders, presenting both opportunities and challenges.
3 *Resources*: The voluntary contribution of time and resources by stakeholders.
4 *Organisation and ground rules*: The establishment of ground rules and organisational structures dictating internal stakeholder behaviour and external relationships.

Conversely, *place-keeping* refers to long-term open space management, including maintenance, partnerships, governance, funding, policy, and evaluation, to ensure that the quality and benefits the place brings can be enjoyed by present and future generations (Dempsey & Burton, 2012; Dempsey, Smith & Burton, 2014). As a newer concept, the case studies highlight additional strategies to consider for application in Singapore (Gopalakrishnan & Chong, 2020):

1 *Accessibility*: The extent of space access facilitating community interaction, physically, socially, and visually.
2 *User Activities*: Space function, including programme type and frequency.
3 *Maintenance/Incivilities*: Space upkeep, including cleanliness, safety, and prevention of misuse.
4 *Level of Autonomy*: The degree of control over space use, maintenance, and management.

To elaborate, the case studies showed that the inception of community initiatives (trigger) often stemmed from interactions among diverse stakeholders. Successful and sustainable initiatives also hinged on partnerships between the community and external stakeholders (actors), including government agencies. These partnerships can be facilitated by events and activities in communal spaces, such as void decks or CCs.

To ensure the longevity and growth of these initiatives, it's crucial to diversify membership through continual recruitment, maintain an open and transparent organisational structure (rules), and rotate leadership roles regularly (actors). Capacity building is necessary for both community members and government officials to foster a productive partnership. Community members need to develop skills in teamwork, leadership, and resource management, while government agencies must learn to engage with and rally community groups effectively, embracing good ideas from all sources without compromising their leadership responsibilities (resources).

The study further revealed that residents and members with direct spatial connection with the shared space (accessibility) will be more effective in generating significant social capital for its long-term sustainability. There is also a need to adopt an innovative approach of participatory governance, with a conscious balance of 'autonomy' and 'authority,' as key to long-term place-keeping (level of autonomy). Efforts that aim to sustain and scale up the initiative such as the regular programming of collaborative events to generate wider interest, create physical and digital visibility, encourage volunteers, build partnerships and networks need to be considered (user activities), in addition to the upkeep of the space to ensure it is clean, organised, and safe for the residents and members (maintenance/incivilities).

Moreover, there is a need for intermediary roles to connect citizens and the government. Although local grassroots organisations like RCs and NCs were tasked to serve this purpose, they may lack the necessary skills and expertise in community design. Community designers can support these intermediaries by enhancing their capacity, facilitating communication, and providing credible information to bolster sustainable community initiatives.

Building upon these insights, together with the new resident segmentation and neighbourhood-based QoL and place analyses detailed in the preceding chapters, data-driven strategies were developed as part of a PAR process. This process aimed to engage residents in three selected towns in Singapore, and to test different models for fostering place-making and place-keeping community initiatives. A three-stage engagement plan comprising co-learning, co-creating, and co-living was hypothesised and test-bedded in each town:

1. Co-learning: Understanding town dynamics and needs through resident and stakeholder engagements, identifying local partners or 'community champions,' and unearthing community assets to develop strategies and tools for fostering social capital among residents.

2. Co-creating: Redesigning and reinventing public spaces through hands-on collaborative processes of place-making, building community capacity, and empowering residents to influence change. It involves the cognitive demands of exerting influence in communities, acquiring the necessary skills and critical knowledge through active participation in community activities (Zimmerman, 1995).

3. Co-living: Sustaining initiatives through place-keeping and integrating them into residents' everyday lives. This requires recognising the power of collaboration and building collective agency amongst the community (Christens, 2012a, 2012b).

## Co-learning

In the first phase of our research, a series of studies were deployed following the PAR approach to gain insights into residents' behaviours within Singapore's public housing neighbourhoods and to solicit feedback on social programmes and community initiatives from both residents and local stakeholders. These studies employed various design ethnographic methods:

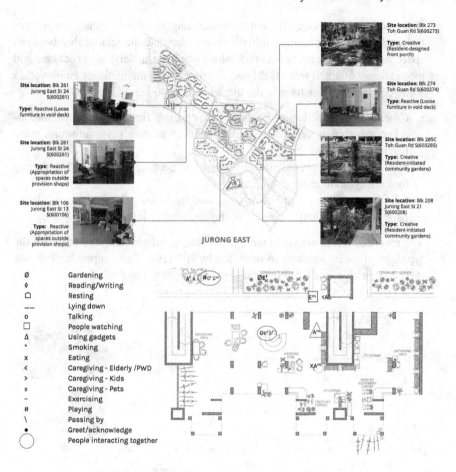

FIGURE 8.2   From top: Overall behaviour mapping of Jurong East; Example of static
sketch of selected void deck.

*Source*: Author.

## Social-Environmental Mapping

- **Asset Mapping:** This involved mapping the spaces, resources, and existing
  community initiatives within the three study sites of Jurong East, Punggol, and
  Toa Payoh. The objective was to comprehend the spatial assets within each
  town, potentially serving as test-bedding sites for future place-making endeavours. The process commenced with cataloguing different spatial assets across
  the selected sites, noting similarities and differences shaped by distinct town
  planning and design strategies. Asset mapping was conducted by researchers
  and residents collaboratively (Figure 8.2).

- **Behavioural Mapping:** This entailed analysing and documenting onsite people activities and community initiatives to understand the relationship between space usage and social behaviours, along with the underlying processes and motivations. Adapted from Gehl and Svarre (2013), the methods employed to study community initiatives in the three towns included:

  - *Counting:* Quantification of activities, ranging from transient to stationary, conducted at various design elements (e.g., pathways, furniture, grass patches, etc.).
  - *Static Sketch:* Representation of activities, people, and places for staying through symbols marked on plans.
  - *Interview:* Utilisation of a semi-structured interview guide to grasp residents' perspectives and understanding of the community initiatives.

The recorded onsite activities were then categorised and analysed based on the typology of social interactions proposed by Hill (2001), encompassing Passive, Active, Reactive, and Creative categories (Figures 8.2 and 8.3).

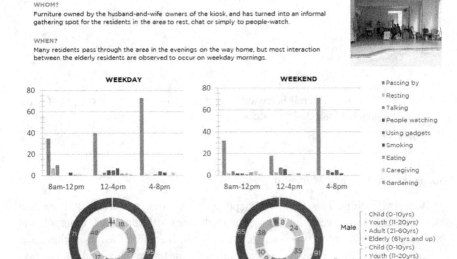

**FIGURE 8.3**    Counting and interview of users at selected case of 'creative appropriation.'

*Source*: Author.

FIGURE 8.4    Left: FGD organised at Jurong East; Centre: Roadshows conducted at three different locations in Toa Payoh; Right: Community Design Bootcamp conducted as part of the Lively Places Programme, an initiative by HDB.

*Source*: Author.

## Community Engagement

Community engagement serves as a vital avenue for fostering collaboration between the research team and residents, facilitating the exchange of first-hand perspectives and insights. Emphasising partnership between the research team and communities, these engagements aim to instigate actionable change (Baum, MacDougall, & Smith, 2006) by uncovering local issues, mapping social capitals, and identifying community champions.

Concurrently with the design ethnographic studies, a series of community engagements (Figure 8.4) were orchestrated in the three towns to establish the research team's presence, cultivate rapport with residents and local stakeholders, and unearth local issues, social capitals, and potential community champions.

- **Focus Group Discussion (FGD):** A medium-scale qualitative participatory approach to delve into the issues faced by participants on the ground. Curated activities and facilitated discussions were employed to collect data. Participants, enrolled through grassroots organisations and local stakeholders, were engaged during multiple time slots, e.g., weekday evening and weekend afternoon, to encompass diverse demographics. FGDs with residents and stakeholders delved into the socio-spatial issues (triggers) within the town, unravelling social capital (resources) and brainstorming potential resolutions, while forging partnerships with like-minded people within the community (actors).
- **Roadshow:** A larger-scale approach designed to engage a broader spectrum of residents who might not actively participate in conventional engagement activities. Conducted as pop-up exhibitions with interactive components across various locations identified through social-environmental mapping and asset mapping, the roadshow amplifies community outreach.
- **Community Design Bootcamp**[3]**:** A medium- to large-scale intensive design thinking workshop aimed at equipping participants with essential knowledge and skills for designing community-centric public spaces. Through engaging activities, participants are introduced to 'participatory analytics,' i.e., where

the research team shares with the participants findings from earlier surveys and mapping that enables them to delve deeper into underlying issues and develop potential projects. The bootcamp fosters ideation, organisation (such as delegating responsibilities and drafting project timeline), and prototyping. It also gave participants the opportunities to consult authorities for advice and plan implementation effectively.

---

**BOX 8.1: WORKSHOP ACTIVITY: PROBLEM TREE**

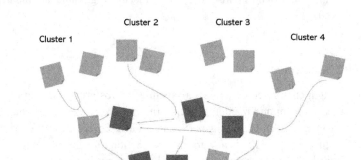

The Problem Tree is a structured approach commonly used for problem-solving during engagements. It provides a graphical representation of the causes of problems, allowing participants to visually organise and connect various causes. This process helps to clarify complex causes, identify contradictions, and ultimately arrive at a shared understanding and objective for finding a robust solution.

Instructions

1 Write down different causes of problems on individual post-it notes, with one cause per note.
2 As a group, organise the post-it notes with related causes into clusters on a Problem Tree board.
3 Draw lines to visually connect different causes written on each post-it note. Use a different colour of post-it notes to represent new ideas, such as summaries or commonalities between ideas, as participants continue to synthesise causes.
4 Synthesise the various ideas and streamline them to form an issue statement as a group. This issue statement will serve as the objective for finding a robust solution.

---

## Co-creating

Building upon the insights and competencies gained through the co-learning process, the communities in the three towns embarked on a journey of community design, actively participating in the co-creation of their neighbourhood spaces. Empowered residents, identified through capacity-building strategies, transitioned into 'community champions,' leading the prototyping of their initiatives in their respective neighbourhoods using digital and place-making tools. Recognising the distinct dynamics of each town, it became evident that a one-size-fits-all approach would not suffice to effectively engage and empower residents from diverse backgrounds. Consequently, a localised strategy of community engagement was developed, tailoring engagement methods to specific localities and accommodating varying levels of participation.

In the following section, we delve into the co-creation processes and outcomes initiated by the residents, beginning with a comparative study of two neighbourhoods: Yuhua in Jurong East and Cascadia in Punggol Waterway.

### *Yuhua, Jurong East: A Case of Asset-Based Development*

Jurong East, home to 79,600 residents, primarily consists of middle-aged individuals and lower middle-class families with adult children. The demographic profile of survey respondents aligns with this composition, with 56% falling between the ages of 40 and 69, 58% residing in three-room or four-room apartment, and 50% earning below the monthly household income of $3,000.

In terms of QoL assessment (Chapter 2), residents of Jurong East express high satisfaction with the Physical and Built Environment (LD3), often describing their neighbourhoods as spacious, tranquil, and well-equipped with a blend of modern and traditional amenities. With many residents having resided in the area for years, a deep sense of belonging is prevalent, scoring above the baseline for Sense of Belonging (LD7). However, areas such as Income (LD1) and Informal Learning (LD5) fall slightly below the baseline (Figure 8.5, top).

The neighbourhood of Yuhua in Jurong East exemplifies similar patterns, comprising an older population with lower formal education levels. The majority of Yuhua residents align with the two primary resident archetypes identified in Jurong East (Secure Homesteader and Silver Contributors, Figure 8.5, bottom), rendering the neighbourhood an apt initial test site within the broader context of Jurong East. Characterised by a slower pace of life, Yuhua residents exhibit emerging interests and passions, seeking local communities and spaces to explore these newfound pursuits. Despite lower formal education levels, there is a notable inclination towards informal learning, indicative of residents' propensity for self-directed learning and community engagement.

Acknowledging the latent skill sets and strong place attachment within Yuhua residents, the research team decided to leverage these assets to enrich the

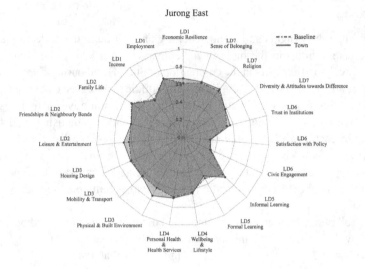

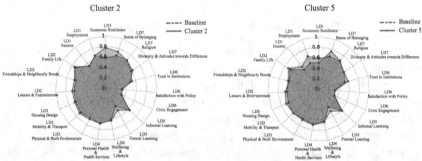

**FIGURE 8.5** Top: Comparison of QoL in Jurong East with Baseline QoL; Bottom: QoL of Resident Archetype 2 Secure Homesteader (left) &and Resident Archetype 5 Silver Contributor (right).

*Source*: Authors of Chapter 2.

neighbourhood. Given the community's satisfaction with the built environment, infrastructural changes were deemed unnecessary. Instead, the focus shifted towards enhancing residents' well-being and fostering social cohesion. Building upon insights from previous design ethnographic studies, potential interventions aimed to cultivate existing social ties to promote lifelong learning, mutual assistance, and potential micro job economies. For instance, initiatives addressing social isolation among the elderly, such as a sharing system, were proposed.

Thus, an assets-based development approach, capable of identifying and harnessing available resources within the community to address specific issues, was embraced. It was imperative to guide residents in nurturing their assets and interests into actionable initiatives, empowering them with the requisite skill sets for community design.

*Project Zero @ Yuhua*

Project Zero marked the initiation of the community design process in Yuhua, facilitated by a partnership between the research team and the local social enterprise, Bold at Work. Based on an assets-based development approach, this capacity-building programme centred on community research and participatory design, aiming to uncover local assets and narratives, foster connections among community champions, and co-generate ideas for community initiatives in Jurong.

Based on discussions generated in previous FGD sessions during the co-learning phase, eight common aspirations for the neighbourhood have been identified in Jurong East:

- Intergenerational and intercultural activity and space
- Barrier-free, healthy, and elderly-friendly environment
- Social inclusiveness for the elderly, children with special needs, and migrant workers
- Cleaner and greener environment
- Improved community infrastructure
- Platform and opportunity for information diffusion and outreach
- Neighbourhood identity fostering through design elements and programmes
- Neighbourhood sharing system to make sharing with neighbours easier

Project Zero facilitated the identification and activation of community champions through a series of outreach initiatives including open calls and empathy field-work.[3] Despite COVID-19 restrictions, the capacity-building programme was conducted through a combination of online and offline platforms, guiding the community champions through a six-part journey to develop and share their ideas with the wider community (Figure 8.6). A total of 12 community champions were identified through the programme. Additionally, Project Zero focused on enhancing the digital literacy and project management skills of these champions, providing guidance to develop and sustain their initiatives.

*Social Deck @ Yuhua*

Project Zero concluded with the development of 'Social Deck,' a transformative place-making kit that reimagines void decks as dynamic social spaces for community champions or organisations to host short-term initiatives. Comprising four adaptable, modular panels, the Social Deck can be easily reconfigured and moved to various void deck spaces in public housing estates. Tested in Yuhua from March to June 2021, the Social Deck facilitated a neighbourhood item exchange and a rescued food sharing initiative, kickstarting a sense of community cohesion. A coalition of resident champions and local organisations, termed 'Friends of Social Deck,' was formed to sustain these activities, fostering participatory governance.

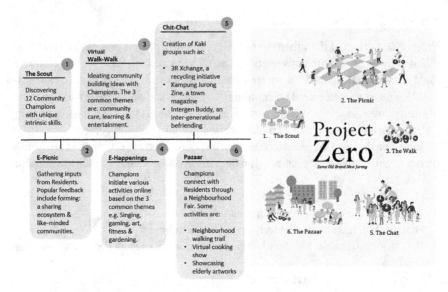

**FIGURE 8.6**   Six-part community champion capacity-building journey in Yuhua, Jurong East.

*Source:* Authors.

The modular design and place-keeping model aim for broader applicability across matured neighbourhoods.

Each of the four Social Deck panels (Figure 8.7) serves a distinct purpose:

- INTRODUCTION: Introduces the Social Deck project.
- FINDING KAKI: Facilitates residents' initiatives, networking, and messages of encouragement.
- JUST KAYPO: Disseminates educational content like recipes and helpful tips.
- PASAR: Hosts a second-hand item exchange corner.

The introduction of Social Deck sparked a heartwarming camaraderie among residents (Figure 8.8). The first pilot, second-hand item exchange corner, 'Pasar,' emerged as the brainchild of Community Champion A, who initially hesitated due to concerns about potential dumping. To address this, Pasar was conceived as a one-way 'gifting' corner for displaying second-hand items in good condition for residents to take. However, within days, residents began contributing their own items and only taking selectively, reflecting the community's strong spirit of sharing. This organic growth of a sharing economy showcased Yuhua's vibrant sense of community, inspiring residents from all walks of life to participate and contribute, further strengthening social bonds.

**FIGURE 8.7**   Social Deck installed at Yuhua, Jurong East.

*Source:* Authors.

**FIGURE 8.8**   From left: Social Deck's pilot—a preloved item exchange corner, 'Pasar'; Second iteration, Food Sharing Point, and its collaboration with the communal kitchen of Kampung Senang; Fourth iteration outside Kampung Senang.

*Source*: Authors.

Encouraged by the success of the first pilot, Social Deck expanded to another housing block within the estate. Community Champion B, inspired by the initiative's impact, saw an opportunity to leverage Social Deck for her Food Sharing Point project, previously piloted using shelving outside her home. Having connected with Community Champion A during Project Zero, she drew inspiration from his experiences in setting up the initial pilot. The Food Sharing Point facilitated the redistribution of imperfect produce from the Fridge Restock Community, ensuring its safe consumption despite minor defects. Initially planned for three weeks, the initiative extended by an extra week due to its popularity, attracting residents from neighbouring blocks to collect produce regularly from Social Deck.

The Food Sharing Point's extensive outreach demonstrated its capacity to connect individuals across diverse backgrounds and make a meaningful impact. Inspired by its success, Community Champion C, who was passionate about cooking healthy food, replicated the programme at her housing block, located near rental flats where residents stood to benefit the most. Meanwhile, Community Champion B noted that residents in her area had initiated their own 'Social Decks,' by placing shelves at the void deck. She has also sustained her food exchange at her void deck, indicating the programme's potential for long-term viability.

Expanding beyond a simple item exchange corner, the Social Deck subsequently evolved into a hub for nurturing meaningful connections. Alongside sharing goods, it fostered a space for heartfelt exchanges, filled with notes of encouragement and personal letters expressing profound appreciation for the community champions. Acting as a spotlight for everyday residents, the Social Deck provided a valuable platform for active engagement with both immediate neighbours and the broader community, weaving a tapestry of interconnectedness and shared goodwill.

### Cascadia, Punggol: A Case of Need-Based Development

Punggol, with its population of 138,700, predominantly comprises young middle-class families. Survey data reveals that 45% fall within the 20–40 age bracket, with 70% residing in four-room or five-room HDB flats and an average household income of $5,000. The dominant resident archetypes, notably CoFam and Multi-Gemners (Chapter 2), mirror the town's youthful demographic profile.

Despite excelling in Income (LD1), Employment (LD1), and Formal Learning (LD5) domains, Punggol falls short in the Living Environment (LD3) domain, being a relatively young town where amenities were still being developed, as highlighted in the QoL assessment (Figure 8.9, top). Interviews underscored residents' desires for enhanced transport, connectivity, and access to key amenities, revealing an opportunity for improvement in Cascadia, a neighbourhood chosen for its representative demographic.

Cascadia's alignment with Punggol's broader resident archetypes makes it an ideal testing ground (CoFam and Multi-Gemners, Figure 8.9, bottom). Its relative newness offers a valuable comparison point with mature areas like Yuhua. The neighbourhood's younger demographic boasts robust knowledge assets, with educated and motivated individuals keen to address infrastructure gaps in their developing estate. Punggol's town planning, oriented towards centralised malls, leaves a void in localised amenities like coffee shops, grocers, and sports facilities, presenting an opportunity for community-driven initiatives to fill these gaps and foster social ties. White spaces that were not yet developed were ideal sites to begin with.

Thus, a needs-based development approach is advocated, empowering residents especially youth and young families, to co-create their community places, bridge amenity gaps, and cultivate a sense of belonging within the burgeoning community.

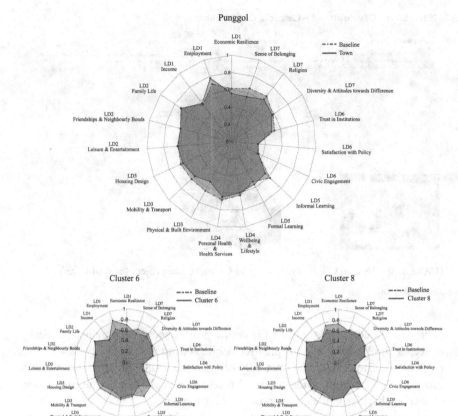

**FIGURE 8.9** Top: Comparison of QoL in Punggol with Baseline QoL; Bottom: QoL of Resident Archetype 6 CoFam (left) and Resident Archetype 8 Multi-Gemners (right).

*Source:* Authors of Chapter 2.

## Design Clinic @ Cascadia

Unlike Yuhua, where the community design process began anew, Cascadia's journey built upon existing ideas from the Cascadia RN. Cascadia Connect, a group of community champions, presented their vision for "Cascadia Our Secret Yard", or COSY[4] to the research team. The chosen site for COSY occupied a vacant area at one end of level 2 of the housing block, left unutilised due to its proximity to the public rooftop garden of the adjacent multi-storey carpark at the same level. Initially overlooked, it held untapped potential for communal purposes. Cascadia Connect sought tailored community engagements to refine their proposal and receive guidance for its realisation.

To facilitate this, a design clinic (Figure 8.10) was conducted. It aimed to guide the COSY project team through a design thinking process, leveraging initial research findings and interactive activities. The clinic enabled participants to: (1)

**Need Based Development – Cascadia Design Clinic**

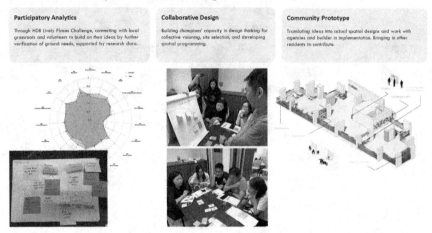

**Participatory Analytics**

Through HDB Lively Places Challenge, connecting with local grassroots and volunteers to build on their ideas by further verification of ground needs, supported by research data.

**Collaborative Design**

Building champions' capacity in design thinking for collective visioning, site selection, and developing spatial programming.

**Community Prototype**

Translating ideas into actual spatial designs and work with agencies and builder in implementation. Bringing in other residents to contribute.

FIGURE 8.10   Design Clinic with Cascadia Connect, the project team of COSY.

*Source:* Authors.

contextualise community needs; (2) define project scope; (3) brainstorm design concepts and features; and (4) plan for public consultation on the design concept (Resetting). The design clinic not only brought the COSY project team closer to ground volunteers and residents, but also served as a platform in the engagement of local authorities in the design process (Matchmaking).

The collaborative efforts culminated in a shared vision to create a "cosy space for residents to explore life in their neighbourhood individually, as a family, or as neighbours," featuring three distinct zones addressing various needs: a co-thinking space, an exploration zone, and a creative playground. Cascadia Connect proposed a more interactive design clinic tailored for the youth in their community, a pivotal moment in the project's progression. By involving a broader range of residents, they created an inclusive platform where diverse perspectives were valued, ensuring COSY's design aligned with collective community needs.

Armed with developed concept plans and visuals, Cascadia Connect shared the intricacies of the COSY project with the Cascadia RN during a dedicated online session. This transparent approach fostered inclusive discussions, allowing stakeholders within the RN to contribute insights. Additionally, it provided a clear articulation of the project's vision and details to local authorities.

## COSY @ Cascadia

The introduction of COSY at Cascadia marks a significant transformation, repurposing an underutilised second-level white space into a dynamic community hub. Furnished since June 2021, it offers residents an inviting space to embrace

**FIGURE 8.11**    Colours of the feature walls co-designed with residents resulted in three distinct spaces in COSY—Commune (left), Collab (centre), and Create (right).

*Source:* Authors.

the essence of neighbourhood life once the COVID-19 pandemic abates, fostering individual exploration, family bonding, and neighbourly interactions beyond their homes. This rejuvenation not only revitalises the once-empty space but also cultivates spontaneous connections and camaraderie among residents, nurturing a stronger sense of community.

COSY comprises three distinct spaces (Figure 8.11):

- *Collab:* A collaborative thinking space outfitted with tables and chairs, ideal for formal activities like studying, working, or conducting workshops, fostering intellectual engagement and collective programmes.
- *Commune:* An activity area merging a pantry with an open space, accommodating various communal activities such as movie screenings and exercise classes, promoting shared experiences among residents.
- *Create:* A creative playground featuring smaller, adaptable spaces for diverse interests like board games, parties, and creative pursuits, encouraging residents to explore their passions.

Each space is visually distinguished by a different colour—blue for Collab, yellow for Commune, and green for Create. To instil a sense of ownership, residents were invited to paint the feature walls in these spaces.

Upon entering COSY, visitors encounter a visually stimulating environment with colourful walls and modular mesh structures (a design language extrapolated from trialled and tested Social Deck), creating an interactive and pleasing atmosphere and setting the tone for a dynamic community space. Beyond its physicality, COSY serves as a curated environment encouraging diverse forms of community engagement and connection.

## Co-living

Co-living emphasises the enduring sustainability and stewardship of programmes and spaces, ensuring their continual alignment with the community's needs and values. The following sections highlight efforts to integrate Social Deck and COSY into the local neighbourhood ecosystem through networking and connecting diverse individuals, fostering new communities to perpetuate these initiatives.

### Yuhua, Jurong East: Integrating into the Local Ecosystem

The inception of Project Zero and Social Deck has fostered extensive connections and synergies, not only within the neighbourhood but also town-wide. A notable collaboration emerged between Community Champion B and Kampung Senang Wellness Studio, a new branch of a charity organisation in Jurong East. Kampung Senang, established in 1999, provides compassionate care to the needy through health-conscious and eco-friendly methods. Their partnership began when the Wellness Studio team collected three days' worth of rescued food from the Food Sharing Point for their vegetarian communal kitchen (Figure 8.8). Recognising the potential to create a sustainable and sharing community in other parts of Jurong East, Social Deck was set up outside Kampung Senang Wellness Studio with the intention of long-term implementation (Figure 8.8). This dynamic collaboration continues to evolve, with Community Champion B playing a key role in supplying surplus fruits and vegetables from the Fridge Restock Community to the Wellness Studio's kitchen. The integration of a nationwide initiative like the Fridge Restock Community into the local context illustrates the far-reaching impact and transformative power of collaboration and community engagement.

To ensure the project's longevity and address potential challenges, the research team led a series of presentations and discussions with various local stakeholders, including Grassroots Leaders (GRLs), Grassroots Advisers (Member of Parliament of Yuhua, Minister of Sustainability Ms. Grace Fu), People's Association (PA), the Jurong-Clementi Town Council, social service organisations including Loving Heart, St. John, St. Luke's, and Tzu Chi, as well as the Singapore Science Centre, which is located right across Yuhua. These sessions served as a platform to share research findings, address concerns, and garner support for Social Deck as a sharing economy enabler. The success of Social Deck resonates with the goals of local GRLs and Grassroots Advisers, leading to a supportive response to further the initiative.

These engagements facilitated the assembly of various stakeholders from the ground up, fostering interconnected relationships within the local ecosystem. Figure 8.12 illustrates the collaborative efforts, where the research team (community designer) worked closely with Bold at Work and Kampung Senang (local community organisers) to empower community champions in developing and running community initiatives (Social Deck). Additionally, external government agencies, organisations, and local residents were engaged to support the initiative, demonstrating the collective effort towards realising a shared vision.

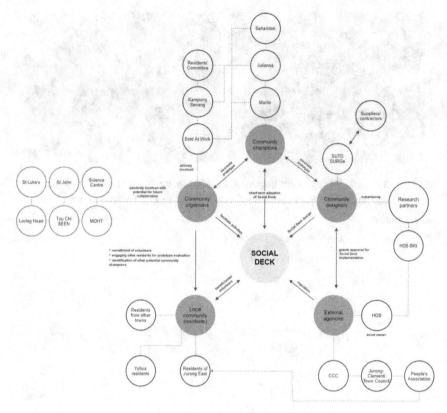

**FIGURE 8.12**  Stakeholder analysis for Social Deck in Yuhua, Jurong East.

*Source*: Authors.

With support garnered from discussions with GRLs, PA, and the Town Council, community champions received approval to conduct their activities with minimal hindrance. Previous studies recognise the potential of adopting flexible partnership arrangements between authorities and citizens to promote transparency and facilitate self-governance in public spaces. The research emphasises the importance of maintaining a conscious balance between 'autonomy' and 'authority' for long-term place-keeping.

## Cascadia, Punggol: Integrating into the Local Ecosystem

Despite the challenges posed by the COVID-19 pandemic, Cascadia Connect diligently engaged residents, particularly focusing on involving the youth, in the conceptualisation of the design. They aimed to sustain the space through both short-term and long-term programmes. Short-term initiatives include events like feature wall painting sessions or space planning sessions to curate the Create space (the creative playground, Figure 8.13, left). Long-term programmes may involve engaging various interest groups to run and maintain initiatives at COSY, such as public talks, recycling

**FIGURE 8.13** Left: Engagement booths at COSY Create space (left); Right: Residents participating in community activities at COSY Commune space.

*Source:* Authors.

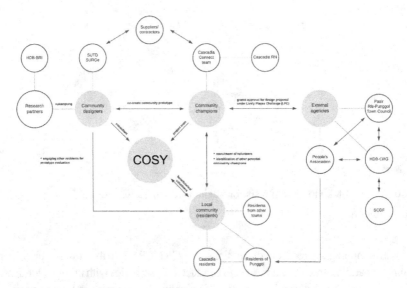

**FIGURE 8.14** Stakeholder analysis for COSY in Cascadia, Punggol.

*Source*: Authors.

programmes, or art and craft sessions (Figure 8.13, right). Involving residents in the final stage of implementation and incorporating their opinions in curating the space can help build a sense of ownership and ensure its sustainability in the long run.

Differences in governance models and neighbourhood demographics between Cascadia and Yuhua have led to the unique stakeholder network relationships (Figure 8.14), which could serve as a model for other younger neighbourhoods similar to Cascadia. In Cascadia, the RN members serve as the community champions (Cascadia Connect), whereas in Yuhua, residents who would usually not be part of grassroots organisations take on the role of community champions. The higher literacy levels of the RN members enable them to independently liaise with

## BOX 8.2: TOA PAYOH—LESSONS LEARNT

Residents in Toa Payoh embarked on a design journey similar to Punggol, initiated from the Community Design Bootcamp as part of the Lively Places Challenge. Care Corner, a social service organisation in Toa Payoh, emerged as one of the community champions during the bootcamp. With extensive experience working with residents, especially the elderly, Care Corner sought design advice and support from the research team to create a community garden. This led to the development of the 'Our Blooming Garden' design prototype (Figure 8.15), intended to foster intergenerational bonding and engage the elderly socially and physically.

However, challenges arose, including construction concerns, a labour shortage exacerbated by the pandemic, and shifts in the organisation leadership and priorities due to the pandemic's impact on seniors' welfare. Regrettably, these factors led to the discontinuation of the project, preventing the prototype's construction. This experience underscores the importance of aligning stakeholder goals, effective communication to address challenges, and the necessity of support and flexibility from authorities to overcome delays and obstacles.

FIGURE 8.15   Our Blooming Garden design prototype developed with Care Corner for Toa Payoh residents.

*Source*: Authors.

other residents and external agencies with minimal support, unlike Yuhua, where Bold at Work and community designers (the research team in this case) actively participate in capacity building and communication facilitation.

## Evaluation of Community Enablement Process and Physical Prototypes

The journey from design ideation to prototype realisation has been enlightening and fulfilling for the communities of both Yuhua and Cascadia. The research team employed a data-driven approach, analysing post-prototype surveys, conducting in-depth interviews with the community, logging prototype usage frequency with sensors, and recording their own observations. This comprehensive approach aimed to identify best practices and challenges in facilitating the dynamic place-making process. The following points highlight the key insights gained from this process: motivation, capacity building, place-keeping, and resource support.

### *Motivation*

Having journeyed with the residents of both Yuhua and Cascadia, it becomes evident that comprehending the needs and motivations of a community is paramount in the place-making process. This understanding highlights the importance of intricately tailoring place-making initiatives to suit each locale. As Yuhua matures, a rooted community driven by individual interests and passions naturally emerges. Residents seek avenues to explore personal inclinations, transforming Yuhua from a mere residential enclave into a dynamic tapestry of interconnected needs and interests.

> I first joined the Facebook group—SG Food Rescue, from there I did rescues in Little India and Pasir Panjang… I decided to start my own food rescue after the Food Rescue organiser encouraged me to do a food sharing point at my place.
> *Community Champion B, Yuhua, Social Deck*

In contrast, in the newly developed town of Punggol, residents are acutely aware of the unmet place-based needs within their community. They perceive these gaps as opportunities and envision the unexplored potential of their surroundings, viewing them as blank spaces waiting to be purposefully utilised.

> … we were quite aware that people may not necessarily want to be as close or even know how to connect. So, one of our guiding objectives was to intentionally design a space that would provide opportunities and facilitate interactions.
> *Community Champion D, Cascadia, COSY*

As a result, the differences in needs are reflected in the responses of each town, with Yuhua focusing more on programmatic aspects, while Cascadia leans towards amenity-based solutions. Collaborating on a community cause that resonates with residents not only equips them with new skills but also strengthens their belief in their role in shaping the future of their neighbourhood. Residents from both towns also emphasised their appreciation for the autonomy granted to facilitate their own discussions, as illustrated by a community champion from Yuhua:

> I don't see that I own it, but I would say that you actually leave it up to us to decide exactly on how we want to do it and we are free to give suggestions of what support we need. So, I will say 100% autonomy, you have given us autonomy on the whole project, and NUK (the research team as Community Designer) has provided whatever that we need.
>
> *Community Champion A, Yuhua, Social Deck*

While residents of Yuhua were actively engaged in the physical setup of Social Deck, Cascadia residents faced limitations due to COVID restrictions during the implementation phase of COSY. One community champion expressed a desire for more hands-on involvement in future initiatives:

> COSY has largely been a project driven by the project team with inputs from residents. I would love to be able to work on a project that leans even more towards co-creation i.e. with more participation and contribution from fellow neighbours in the making of a space like this. Perhaps this can be considered for phase 2 or 3, or for another project?
>
> *Community Champion E, Cascadia, COSY*

Clearly, by understanding the unique needs and motivations of diverse communities and granting them the necessary trust and freedom to explore their aspirations, a nuanced and contextually relevant approach to place-making can be developed. This fosters a more authentic and resonant environment for the individuals who call these places home.

Analysis of post-implementation survey data revealed a significant correlation between the prototypes and increased interest among residents in community activities. In Yuhua, the Pasar (item exchange corner) was well-received, with residents acknowledging Social Deck's role in raising awareness about neighbourhood activities and contributing to a stronger community spirit. Meanwhile, residents of Cascadia appreciated COSY for its spaciousness, comfort, and restorative, welcoming atmosphere. Visitors noted that COSY facilitated connections with neighbours and sparked interest in future community involvement. While these findings offer valuable insights, further studies with a broader population would enhance their applicability. Nonetheless, these results underscore the efficacy of both prototypes as catalysts for future place-making initiatives, encouraging residents to actively engage in their community.

*Capacity Building*

In the realm of community engagement, the community champion emerges as a central figure deeply involved in prototyping and ideation. However, despite their sincere efforts, obstacles often arise, particularly due to a lack of design or technical expertise and difficulties in obtaining regulatory approvals. These challenges not only impede progress but also serve as discouraging factors, hindering other residents from actively participating in collaborative initiatives. This highlights the urgent need for support from community design expertise.

> (Authorities) have their own new processes, which we have to follow quite strictly. And so that if because it's a community project, I felt that they (authorities) can give a bit of free play, then I think it will be easier to execute, you know, because especially because we are community (without professional building experience), we are doing this on a volunteer basis.
>
> *Community Champion E, Cascadia, COSY*

Faced with these challenges, a distinctive strategy unfolded, exemplified by the context of Jurong East. Here, the role of a community designer took centre stage, introducing a fresh dynamic to the community engagement framework. This innovative position leveraged expertise and resources to provide design and technical guidance to the community champions, capitalising on the inherent capacity of the community. Community designers, finely attuned to the immediate needs of the residents, played a crucial role in refining and materialising ideas, seamlessly bridging the gap between community aspirations and regulatory requirements. This vital supportive role not only facilitated smoother collaboration between residents and authorities but also empowered the community to bring their visionary concepts to fruition in a more efficient and cohesive manner.

> I think that the change is the progress that I have always wanted to make. Because when I do it on a personal basis the progress is very slow. The outreach is not there. But after you all (Community Designers) came into the whole thing; it definitely has helped to move so much further.
>
> *Community Champion B, Yuhua, Social Deck*

> Through the partnership with SUTD, I have come to understand and learn more about the elements that go into designing good community spaces. Other than some of the more visible projects, to the (design) untrained eye it is easy to take many of the design features for granted. I also think that design should be a baseline requirement for future new community spaces, both pre and post build.
>
> *Community Champion, Cascadia D, COSY*

## *Importance of Place-Keeping*

Amidst the towns, diverse visions take shape as residents formulate strategies for the sustained vitality of their initiatives. Particularly noteworthy is the effort in Cascadia, where community champions are leading intentional campaigns to spread awareness about the community prototype, with a special focus on engaging the younger demographic. Beyond addressing immediate needs, these champions see the prototype as a valuable asset for the youth, actively seeking their continuous involvement for the future. This strategic approach entails nurturing a sense of community engagement and ownership among the younger population, acknowledging their significance as stakeholders in the ongoing story of the community's evolution.

Furthermore, community champions are strategically expanding their horizons beyond the boundaries of the community itself and are actively seeking collaboration with external entities, particularly community partners such as nongovernmental organisations (NGOs). This collaborative effort aims to secure additional support in terms of manpower and caretaking. By forming partnerships with these external stakeholders, the resident champions aim to enhance the capacity and resources available for the sustained success of their initiatives, cultivating a broader network of support that extends beyond the immediate confines of the residential areas. This collaborative and forward-thinking approach highlights the residents' dedication not only to initiating but also to nurturing and sustaining community-driven projects for the benefit of both present and future generations.

> I would think collaborating with NEA and organisations with active youth volunteers (for example, from the tertiary schools) would have boosted some of our efforts.
> *Community Champion E, Cascadia, COSY*

> I think if we (Social Deck) can move to more places (in Jurong East) that'd be great. I'm sure the RCs will support and they will love to have that. They can help to check the quality of the items. Maybe every Monday as a one-week program, or maybe on a quarterly basis, we get this event for the RC.
> *Community Champion C, Yuhua, Social Deck*

In discussing the sustainability of community initiatives, community champions emphasised the importance of both hardware, such as maintenance and upkeep, and software aspects, like programming. In Yuhua, two of the initiatives focused on establishing regulations to prevent residents from misusing the system, such as hoarding items in bulk or treating the area as a waste disposal site. Community champions shared the rules for initiatives, highlighting the necessity of clear guidelines to ensure the initiative's longevity and effectiveness.

> Take what you need, leave what you don't need; don't have ugly behaviour—taking items in bulk i.e. instant noodles, biscuits, rare items, pop tarts, canned

food, usually food that has a longer shelf-life people like to take immediately and in bulk.

*Community Champion B, Yuhua, Social Deck*

Whatever that somebody wants to give away, it must be in good working order. So, if it is an electronic item, it can be old, but if it is still working it is fine. As for the clothing, as long as it's not too faded, then it is acceptable.

*Community Champion A, Yuhua, Social Deck*

In the case of COSY, the initiative challenged cultural norms by welcoming everyone to access the open concept community space, aiming to foster a sense of ownership and camaraderie among residents. However, this approach requires extra caution in maintaining the space to prevent potential incidents such as damage or misuse.

But this, we're trying out a totally different model, where we're leaving things out for people to self-service and all. In return, we are hoping that they will reciprocate that and then take care of the equipment. But we already are experiencing a situation where residents play with the fire extinguisher and things like that. So, I think that's one of the challenges that we face, especially at the start of the project. So, can you imagine like a month go by, the maintenance, the upkeep of the space, it's going to be something that we need to look into as well. Which is above and beyond what the challenge is currently.

*Community Champion D, Cascadia, COSY*

### Resource Support

Adequate resources are essential for both the establishment and the maintenance of community initiatives. In Yuhua, the Social Deck programme initially relied on connections to various stakeholders and operated on a voluntary basis. Consequently, most community champions utilised existing resources and funded the initiatives from their own pockets. However, this approach inevitably constrained the scope and scale of their efforts. Despite the abundance of materials for recycling and waste reduction initiatives, there was a shortage of manpower to effectively manage them, as highlighted by community champions.

My ideal structure would be to have volunteers and drivers besides just myself collecting and distributing it. (I) also need money for petrol and also not everyone is willing to volunteer. Volunteers need to help from start to end, including cleaning up, setting up the whole free market, volunteers also have the privilege to take food items first, so I need people who are not volunteering just because of that.

*Community Champion B, Yuhua, Social Deck*

For Social Deck: I will drop by around 4 pm every day, rearrange the items, get rid of unsuitable items with stains etc., update the WhatsApp chats with photos so people know what is available. For the Weekly Food Pasar: (I) get food supply from Fridge Restock Community and run the free pasar from 1 pm ~ 4 pm every Tuesday, and also have leftover items outside my home and will update the group chat whenever there are new items. Even the cleaners do not throw the items away, they will clean the area and put any items back to the original place.

*Community Champion A, Yuhua, Social Deck*

Cascadia, on the other hand, faced budget limitations. Despite having the Lively Place Challenge grant, residents wrestled between financial decisions and fulfilling their design aspirations for COSY, as they quickly realised the high cost of labour and building materials.

But even just merely trying to lay the ground, you know, trying to paint the walls and all that, all these all add up, electrical work was already a few thousand already. So, it's like, they just chew away at the budget quite quickly.

*Community Champion, Cascadia E, COSY*

Because really, it's quite a small quantum for the level of work that we wanted to do. Just the floor already was like $6000-$7000. So yeah, we didn't even; we couldn't even think about doing any tiling or anything like that.

*Community Champion D, Cascadia, COSY*

Therefore, having the right and substantial resources such as manpower and financial aid would greatly benefit future place-making and place-keeping endeavours, allowing the residents to transform their creative visions into tangible, impactful projects, helping them to support their communities.

Drawing from our experience collaborating with communities in Jurong East, Punggol, and Toa Payoh, it becomes evident that the dedication of community champions, alongside the support of grassroots organisations and external agencies, significantly influences the success of community prototypes in each locality. The community champions across these areas exhibit diverse profiles: those in Jurong East pursue initiatives aligned solely with their personal interests, while Punggol's champions form a committee representing residents' interests. Conversely, community champions in Toa Payoh act as representatives of their clientele, prioritising their clients' needs over personal or communal interests. This variation in profiles directly impacts their commitment levels.

In Jurong East and Punggol, community champions display remarkable dedication, willingly investing personal time beyond their working hours to implement community prototypes. Their enthusiasm and commitment are palpable as they devote extra hours, reflecting a strong passion for their initiatives. Conversely, the community champions in Toa Payoh, though equally devoted, faced challenges balancing prototype tasks with their regular workload during working hours. This insight underscores the necessity of adapting community design strategies to accommodate the varying capacities and interests of local stakeholders (Melcher, 2013).

The presence and support of grassroots organisations and external agencies differed significantly across the three towns. In Yuhua, community champions benefited from introductions to various community organisers, which expanded their networks and streamlined the approval process for initiatives. This collaborative environment fostered dynamic momentum, with regular updates and progress. Conversely, Toa Payoh encountered challenges due to a lack of response from local stakeholders, resulting in occasional delays. This highlights the importance of support from local organisations, as they can either facilitate or hinder place-making efforts, despite strong support from agencies at the national level.

## Community Enablement Framework

The implementation of the PAR approach across the three towns provided a nuanced understanding of each town's distinct requirements. The place-making process served as a valuable platform for equipping various community champions with essential intangible skills vital for personal growth and community betterment, including teamwork, creative problem-solving, and adaptability. For many participants, this marked their initial foray into actively contributing to their neighbourhood's management. Drawing from these experiences, the research team has developed a Community Enablement Framework (Figure 8.16), serving as a blueprint to enhance collaborative relationships among residents, community stakeholders, and agencies.

To facilitate future community enablement within local towns, a five-stage action plan is proposed for reference by future town planners and community designers:

**Scouting:** Identifying Champions and Assets
Community designers identify community champions and their skills through workshops, engaging with residents, fostering community-driven initiatives in later stages.
**Incubating:** Activating Champions, Building Capacity and Guiding Development.
Community designers mentor community champions in understanding community needs, utilising their skills, and equipping them with the required knowledge and skills to plan and organise their ground-up initiatives.
**Resetting:** Collaborative Prototyping.
Community designers guide community champions in implementing and testing ground-up initiatives, gaining insights on resident responses, ensuring sustainability of the project.
**Matchmaking:** Integrating into the Local Ecosystem
Community designers connect community champions with local authorities and organisations, to ensure alignment with existing networks, protocols and goals specific to each town to build a sustainable community ecosystem.

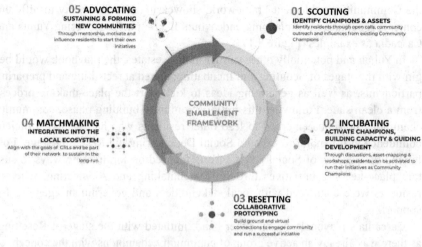

**FIGURE 8.16**   Community enablement framework.

*Source*: Authors.

**Advocating:** Sustaining and Forming of New Communities.

Existing community champions are encouraged to mentor other residents in their journey to become community champions. By facilitating the empowerment of others through asset identification and strategic insights, they contribute to a self-sustaining model for community enablement (Christens, 2012a, 2012b).

Previous understanding of residents' motivations has established that the place-making process must be location-specific. We demonstrate the flexibility of the Community Enablement Framework, showcasing its adaptability to different contexts for future town planning endeavours like a 'playbook,' using Yuhua and Cascadia as examples (Figure 8.17).

In Yuhua and potentially other similar mature estates, the playbook would begin with the stages of Scouting and Incubating, aimed at recruiting and preparing participants, as well as generating ideas to kick-start the place-making process from a clean slate. Following this initial community-building phase, community champions in Yuhua emerged to lead other residents who had gained sufficient confidence at this point to execute Social Deck during the Resetting stage. Following the success of Social Deck, efforts focused on formulating ways to sustain place-keeping initiatives through Matchmaking and Advocating, wherein residents were connected with local stakeholders and government agencies for support.

Cascadia's playbook, on the other hand, initiated with the stage of Resetting, as there was already an active group of community champions with the concept of COSY in mind that needed assistance in execution. Matchmaking and Incubating were then carried out, primarily aiming to connect residents with local stakeholders for collaboration and advice, as well as to equip residents with technical building knowledge and skills. After COSY was completed, Scouting and Advocating took place to expand the pool of community volunteers and foster partnerships with the government for future initiatives. This scenario could potentially be applied in other young neighbourhoods.

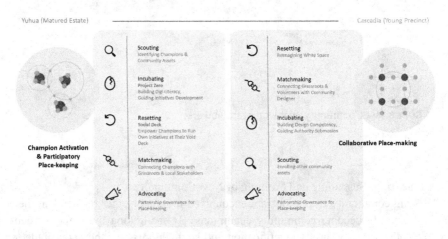

**FIGURE 8.17** Community Enablement Playbook adapted from the framework: Yuhua, Jurong East versus Cascadia, Punggol.

*Source:* Author.

## Conclusion

This study introduces a neighbourhood-specific community engagement model and diverse engagement strategies employed by the research team across three towns, organised within the five-stage action plan of the Community Enablement Framework that serves as a playbook. The playbook, along with accompanying engagement strategies and toolkits,[5] serves as a reference for town planners, community designers, social service providers, and all other community organisers. Its uniqueness lies in its emphasis on identifying appropriate points of entry and capacity building.

To initiate the community enablement process, identifying the right point of entry is crucial, in terms of town demographic, QoL profiles, predominant resident archetypes, residents' digital literacy, etc., resulting in the needs of specific engagement strategies for each town that oftentimes require expert knowledge and intuition. Traditionally, local institutions or authorities like PA and Town Council serve as entry points, but the study reveals that this is not always the case due to misalignments in agendas, especially during unprecedented times like the COVID-19 pandemic. Alternatives must be explored when traditional avenues yield lukewarm responses. Acknowledging the need for capacity building among residents, local authorities should encourage their active involvement, necessitating an inter-governmental effort to streamline processes and bring about self-enabled communities in HDB neighbourhoods and towns.

While this framework outlines key stages in community enablement, it is cautioned against being considered a one-size-fits-all solution for other neighbourhoods and towns due to differing dynamics, necessitating specific engagement strategies that often require expert knowledge in deriving insights from data and past experience. As such, the organic nature of community engagement requires continuous effort and meticulous planning based on expert insights, making the role of community designers indispensable. Community designers are crucial in scaffolding residents through the extensive administrative processes that many community-initiative programmes, and funds from various government agencies, require. Faced with this challenge with no prior experience or expertise in the field, residents are otherwise deterred by the lengthy processes. Besides that, the subsequent stages of the Community Enablement Playbook focus on residents' talents and assets, leveraging their capabilities, making them feel valued, encouraging service-oriented drives, prototyping ideas, and providing logistical support, all requiring professional resources beyond administrative support, which further underscores the crucial role of community designers.

Reflecting on the varying dynamics across the three towns, Jurong East, Punggol, and Toa Payoh, the study has not only shed light on the various driving forces behind successful place-making but also inspired a hopeful outlook. It is with optimism that the insights garnered from this research serve as a catalyst for advocating continuous place-making endeavours, emphasising the imperative of meticulous planning and collaborative initiatives. By understanding the intricacies of each town's and neighbourhood's unique context, followed by collaborative approaches and sustained efforts, vibrant and resilient communities that authentically reflect the needs, aspirations, and identities of the diverse residents within these urban living spaces can be fully realised.

## Acknowledgement

The authors express their sincere gratitude to the grassroots advisers, grassroots organisations, town councils, People's Association, local stakeholders, community champions, and all residents in Jurong East, Punggol, and Toa Payoh who actively participated in the community engagements during the research period. Special thanks are extended to Bold at Work, Loving Heart, Kampung Senang, and the Singapore Science Centre in Jurong East, the Cascadia RN in Punggol, and Care Corner in Toa Payoh for their steadfast support, especially during the challenging times of the pandemic.

## Notes

1 The history of CC can be traced back to 1948, and later on its name changed to Community Club. RC was introduced in 1978 for public housing estates, while the NC was established in 1998 for private housing estates. RN was set up more recently in 2018 to combine and dissolve the boundaries of RC and RN and to promote social mixing between the estates. See https://www.pa.gov.sg/our-network/grassroots-organisations/grassroots-organisations/

2 PAR is an approach combining research, education, and action used to gather information to inform decisions made concerning social or environmental issues. In PAR, researchers work closely with members of the target community, and tap on a variety of skill sets to engage with people and draw insight from the process. See Kindon, Pain & Kesby (2007).

3 Open calls include word of mouth, pop-up booths, social media campaigns. Empathy fieldwork includes interviews, narrative inquiry meet-ups.

4 Residents from Punggol and Seng Kang were recruited through the Lively Places Challenge 2019, a public housing initiative which encourages communities to initiate community bonding projects in neighbourhood public spaces. In Cascadia, participating groups included residents and grassroot leaders who formed the Cascadia Connect, the project team for COSY.

5 For more community engagement strategies and toolkits, refer to www.soulab.co.

## Bibliography

Baum, F., MacDougall, C., & Smith, D. (2006). Participatory action research. *Journal of Epidemiology and Community Health, 60*(10), 854–857. https://doi.org/10.1136/jech.2004.028662. PMID: 16973531; PMCID: PMC2566051.

Chong, K., Gao, T., & Kato, Y. (2019). Heartland Kaki: Creative appropriation and community design in public housing. In *Proceedings of the 12th Conference of International*

Federation of Urbanism (IFoU) 2019: Beyond Resilience – Towards a More Integrated and Inclusive Urban Design, Jakarta, Indonesia, June 24–26, 2019.

Christens, B. D. (2012a). Targeting empowerment in community development: A community psychology approach to enhancing local power and well-being. Community Development Journal, 47, 538–554. https://doi.org/10.1093/cdj/bss031.

Christens, B. D. (2012b). Toward relational empowerment. American Journal of Community Psychology, 50(1–2), 114–128. https://doi.org/10.1007/s10464-011-9483-5

Dempsey, N., & Burton, M. (2012). Defining place-keeping: The long-term management of public spaces. Urban Forestry & Urban Greening, 11(1), 11–20.

Dempsey, N., Smith, H., & Burton, M. (Eds.) (2014). Place-keeping: Open space management in practice (1st ed.). Routledge. https://doi.org/10.4324/9780203725313

Gehl, J., & Svarre, B. (2013). How to study public life. https://doi.org/10.5822/978-1-61091-525-0

Gopalakrishnan, S., & Chong, K. H. (2020). The prospect of community-led place-keeping as urban commons in public residential estates in Singapore. Journal of Built Environment, 46(1), 100–123.

Hill, J. (2001). The use of architects. Urban Studies, 38(2), 351–365. https://doi.org/10.1080/00420980123765

Holton, J. (2007). The coding process and its challenges. In The SAGE handbook of grounded theory (pp. 265–289). SAGE Publications Ltd. https://doi.org/10.4135/9781848607941

Melcher, K. (2013). Equity, empowerment, or participation: Prioritizing goals in community design. Landscape Journal: Design, Planning, and Management of the Land, 32, 167–182. https://doi.org/10.3368/lj.32.2.167

Kindon, S., Pain, R., & Kesby, M. (2007). Participatory action research approaches and methods: Connecting people, participation and place. Routledge.

Project for Public Spaces (n.d.). What is placemaking. Retrieved from https://www.pps.org/article/what-is-placemaking

Zimmerman, M. A. (1995). Psychological empowerment: Issues and illustrations. American Journal of Community Psychology, 23(5), 581–599.

# 9

# BRIDGING PHYSICAL AND VIRTUAL COMMUNITY THROUGH O2O APPROACH

*Keng Hua Chong, Denise Nicole Lim, Natasha Yeo Min, Tshui Mum Ha and Sin Mei Cheng*

## Introduction

Amidst unprecedented times, a sliver of hope is found in the plethora of community initiatives that have steadily emerged in response to the global pandemic crisis. In Singapore, many of these are self-organised initiatives, often treading the online sphere in the spirit of mutual aid and solidarity, garnering public assistance to provide support to fellow citizens who have slipped through the cracks of government response. The rise of crowdsourced aid is reminiscent of Deputy Prime Minister Heng Swee Keat's 2019 dialogue, echoing the sentiments of former foreign minister S. Rajaratnam who coined the term 'democracy of deeds':

> This is what I mean by a democracy of deeds – contributing not just ideas but also effort. I am confident that many new and exciting ideas, and many constructive actions, will surface. As long as we persist, learn from each other, we can forge a new way forward, step by step.
>
> *(DPM Heng Swee Keat, 2019)*

The challenges of COVID-19 have not only reignited the spirit of 'democracy of deeds' but necessitated it. While effective in curbing the spread of the virus, measures such as 'Home-based Learning' and issuances of 'Stay Home Notice(s)' prove challenging to vulnerable groups that have limited resources and are therefore ill-equipped for an extended period of isolation. Yet, these issues are neither new nor surprising, with many stemming from deep-seated problems that have long existed prior to COVID-19: low-income households struggling to make ends meet, elderly living alone, overcrowded lodging of foreign workers, etc. The impact of the pandemic has amplified their plight, making them visible in ways that call for immediate action and response.

DOI: 10.4324/9781003437659-13

Beyond offering aid, many have also taken the period of isolation as an opportunity to share knowledge and care through virtual workshops and webinars; messages of solidarity with frontline workers are often seen circulating on social media feeds; home-based businesses have surfaced as many are finding means to sustain themselves through the economic crisis – the coronavirus has taught the world how to live online.

In Singapore, a sharp increase by more than 250,000 internet users was observed in mid-2020 (Kemp, 2020), coinciding with the peak of the pandemic. Today, 99% of resident households have access to the internet (IMDA, 2023), making Singaporeans no stranger to a digitised life. Many have explored creative ways to use digital tools to compensate for limitations on physical interactions. While there are concerns about a widening digital divide that need to be addressed on a systemic level, with the support from crowdsourced efforts garnered through digital tools, volunteers on the ground have been working hard to fill the gaps by assisting these "digital outcasts" (Ong, 2020). Technology thus becomes an enabler that can equip the public with data to efficiently provide open-source assistance based on proximity and priority, effectively redistributing resources to areas in deficit. There is strength in numbers: leveraging on collective efforts can help to ease the burdens of those treading water during this pressing time.

The rise of virtual community initiatives is a phenomenon worth exploring in the unparalleled scenario of a global lockdown, to provide insights into how people self-organise using digital means. By taking stock of recent virtual community initiatives in Singapore, the following case studies aim to establish a new framework for virtual community-led urbanism, with emphasis on: (1) the crowdsourced dynamic of virtual community initiatives, and (2) understanding how O2O[1] . (offline-to-online/online-to-offline) methods have facilitated such community-driven activities amidst social distancing. This will serve as a reference for a new model of urbanism that necessitates the hybridisation of physical and virtual networks and tools, to build a more collaborative community and shared future. The case studies culminated in the development and evaluation of a community-driven online application 'nuKampung', rolled out in conjunction with a series of engagements on ground. The objective was to witness the transition of activities between the physical realm and the online domain, therefore marking the possibility of O2O hybridisation as a novel way of involvement for future community engagements.

## Case Studies

The case studies were chosen based on the shared characteristic of users engaging in activities that encompass both online and offline dimensions, requiring operational transitions between these two realms. For example, case study 'Belanja Eat' provides information online for those in need who wish to be blessed with a free meal at participating food establishments, before visiting these eateries which will then physically prepare the meals onsite for them. Both background aspects of the applications, such as operational strategies, and foreground aspects, such as usability, were studied extensively.

In the study of operational strategies, 11 virtual community-led initiatives that emerged during the 'Circuit Breaker'[2] period in Singapore were chosen for analysis (Table 9.1), with the aim of uncovering methods creatively employed by these initiatives to sustain their operations in a time of limited contact and scarce resources. The relationship between the virtual initiatives and their social-spatial contexts is clarified using process mapping, a research method that facilitates the identification and evaluation of key stages of a project (Jupp, 2006). The initiatives that facilitate the flow of resources from an area of excess to that of deficit can thus be generalised into five stages of operation: (1) Gathering of resources/needs, (2) Preparation

**TABLE 9.1** Case studies for operational strategies

| No. | Selected Case Studies | Description of Operational Strategies |
| --- | --- | --- |
| 1 | Computers against Covid (CAC) | Collection and refurbishment of old laptops for students on 'Home-Based Learning' |
| 2 | SG Makers against Covid | Crowdsourced 3D printing of ear savers for distribution to hospitals |
| 3 | #BraveHeartSG | Campaign that encourages sharing messages of solidarity with frontline workers using the hashtag #BraveHeartSG; notes are printed out, packed, and distributed to hospitals |
| 4 | Belanja Eat | Pay-it-forward meal system at participating hawker stalls |
| 5 | Wares Mutual Aid | Solidary spreadsheet that lists needs and resources for the public to use as a directory to provide mutual aid |
| 6 | The Good Space | Solidarity spreadsheet listing initiatives and projects currently active; provides a platform for initiatives to share their process and call for resources if necessary |
| 7 | Mask Go Share | A website that maps one-bedroom and two-bedroom rental flats for the public to leave masks and other essentials in their mailbox; also lists needs and resources |
| 8 | Scratchbac | Telegram Bot that notifies users if they are within a radius (200 m to 1 km) of a request |
| 9 | Hawkers United | Facebook group that hosts lists of hawkers that provide delivery service during the 'Circuit Breaker' period |
| 10 | Aid Hub | Website that consolidates initiatives in Singapore |
| 11 | I Am a Community Care Buddy (CCB) | Website that has resources ranging from mask collection points to job hunting resources to help the public tide through the 'Circuit Breaker' |

*Source*: Authors.

of resources, (3) Matching of resources with beneficiaries, (4) Distribution of resources to beneficiaries, and, (5) Advocacy of the initiative. The research team delved into these five different stages of operations, to better understand ways of streamlining processes, managing resources, and improving efficiency.

As seen in the emergence of crowdsourced efforts organised using informal means (such as social media, Google sheets, and other online platforms), stakeholders, rules, and activities are no longer clearly defined, but are instead often found to be a collaborative effort between initiators and the general population who have stepped up to provide mutual aid. Thus, it is valuable to understand the spectrum and types of crowdsourced collaborations, as well as how online platforms have facilitated this phenomenon. As the scenarios presented in crowdsourced initiatives are often complex and hard to quantify due to the unpredictability and shifting nature of such an open system, a rubric system is introduced to provide a system of objective assessment with detailed descriptions of clearly discernible attributes (Salkind, 2010). Here, the designated rubrics are represented using symbols with assigned values (+1, −1) to indicate the extent of crowdsourced versus self-managed organisation and online versus offline strategies, while also accounting for a hybrid case of each extreme (±0.5).

Patterns and commonalities emerged after comparative processing mapping (Figure 9.1) was conducted across the five stages of operations of each case study. The case studies were then categorised according to their intensity of

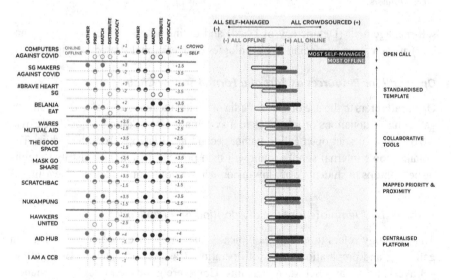

**FIGURE 9.1**   (Left) Summarised process maps; (right) stacked bar chart indicating aggregated proportions of crowdsourced/self-managed organisation and use of online/offline platforms.

*Source*: Authors.

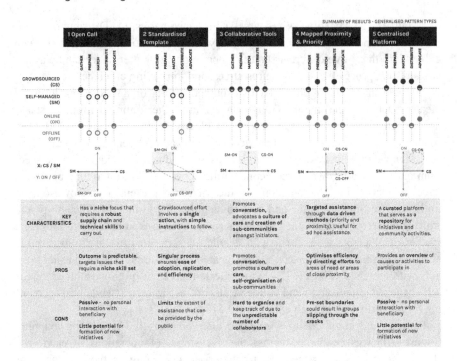

**FIGURE 9.2** Five typologies of operational strategies.

*Source*: Authors.

self-management (Democracy of Deed) and online activity (Technology as Enabler) at each stage, surfacing five typologies of operational strategies (Figure 9.2).

### Open Call for Resources as Part of a Formal Supply Chain

Open call refers to the use of social media or digital means to appeal to the public for gathering of donations or resources in a systematic way. Computers against Covid (CAC) organises an open call to gather second-hand laptops from the public. The organisation's internal supply chain will then prepare, match, and distribute refurbished laptops to students from low-income households for home-based learning.

### Standardised Template for Ease of Adoption and Replication

This typology refers to initiatives typically supported by crowdsourced efforts in gathering and preparation stages of operation. Case studies such as Belanja Eat, #BraveHeartSG, and SG Makers against Covid are good examples. For instance, the #BraveHeartSG movement garners crowdsourced support through the collection of messages of solidarity with frontline workers. The public are encouraged to share their messages on social media with the hashtag #BraveHeartSG, allowing it to be searchable for consolidation, printing, and distribution to hospitals.

### Coordination of Resources and Facilitation of Open Discussions Using Collaborative Tools

The collaborative use of open-source tools, such as Google sheets, has gained significance in wake of the pandemic. For example, the Good Space's Live Updates offers a platform for the listing of initiatives with the aim of helping projects in need. A key observation of such methods is that, while facilitating communication across initiatives, they allow for open conversation and discussion amongst community initiatives as well as with the public.

This sentiment of care is also apparent in Wares Mutual Aid, emphasising the need for sharing, also encouraging people to form their own clusters in the process.

### Mapped Priority and Proximity for Targeted Assistance

Initiatives of this typology use digital tools to facilitate targeted assistance through priority mapping and geotagging. Process evaluation case studies such as Mask Go Share (priority mapping) and Scratchbac (proximity mapping) exhibit that this typology supported crowdsourced efforts in the gathering and matching stages of operation, whereas the preparation and distribution stages were independent efforts directly between donors and beneficiaries.

### Centralised Platform Hosting Consolidated and Curated Content

The final typology focuses on gathering of resources using submission platforms that are vetted by the initiators, creating a curated and safe space for the public to browse and contribute to the causes they are most drawn to. Hawkers United, Dabao 2020 consolidates hawker stalls, which under normal circumstances rely heavily on walk-ins for businesses. This provides visibility for such stalls and creates convenience in browsing a single page to find hawker food delivery amidst social distancing measures.

Other initiatives such as AidHub.sg and I Am a Community Care Buddy (CCB) consolidate helpful tips and services from medical needs, volunteering opportunities, to news updates, serving as a consolidated platform to help the public tide through the 'Circuit Breaker' period.

The five typologies outlined – open calls for resources, standardised templates, collaborative tools, mapped priority and proximity, and centralised platforms – offer a comprehensive framework for enhancing and optimising resource mobilisation during times of crisis. Embracing open calls can efficiently gather and distribute resources. Standardised templates illustrate the potential for replicability and widespread adoption. The coordination of resources through collaborative tools fosters open discussions and strengthens community bonds, while strategic use of priority and proximity mapping ensures targeted assistance. Finally, centralised platforms provide curated spaces for streamlined information dissemination. Keeping in mind the integration of these approaches opens the possibilities of developing future community-based applications that foster a more resilient and interconnected response to future challenges.

Further investigation into the user experiences of online applications also uncovered the adoption of gamification as a strategy to enhance both user engagement and retention. Gamification involves the integration of game elements and mechanics such as competition, point systems, rewards, and challenges to enhance user engagement, making experiences more enjoyable and compelling. Diverse forms of online interaction were observed, including text, images, videos, and link sharing, contributing to the multifaceted nature of online engagement. The incorporation of reward and point systems, along with a light-hearted interface, creates engaging activities that encourage sustained involvement. Notably, the online features are updated to the local context, integrating familiar elements like colloquial language or places, thereby creating an online environment that resonates with the community's preferences and behaviours. Interesting online gamification behaviours such as the presence of a virtual community reveal a fascination with collecting items, evident in crazes like Hello Kitty and PokemonGo. Other common behaviours observed reflect a preference for community regulation over top-down governance, which is characterised by a distinctive blend of unity and critique. There is a tendency for individuals to rally together, expressing their opinions and concerns collectively, reflecting a vocal and opinionated user base. Simultaneously, the community exhibits a penchant for oversharing and interaction, as seen in the widespread use of memes.

Understanding online application features and user behaviour enabled the research team to comprehend typical online-offline usage patterns. These patterns are summarised in an 'O2O experience' framework below, comprising the following five stages:

- **Consumption:** Passively observe traffic and activities online
- **Interaction:** Ask and learn from the community
- **Contribution:** Commenting on posts and connecting with other users
- **Collaboration:** Making friends online and forming communities
- **Initiation:** Starting and organising events for the rest of the community

The understanding of the five typologies of O2O operational strategies and five stages of O2O experience helped the research team to develop an experimental O2O platform to enhance neighbourly interactions and promote community initiatives, complementing the place-making and place-keeping efforts outlined in Chapter 8.

## nukampung: A Pilot Community Mobile App

Being cognizant of key insights gathered from the case studies, the research team embarked on the task of developing a community-based mobile application that aligns with and enhances various activities planned for three local towns in

Singapore: Jurong East, Punggol, and Toa Payoh (see Chapter 8). The outcome is the 'nukampung',[3] a community lifestyle application that serves as a one-stop platform for all the events and activities within a neighbourhood. It aims to connect residents with local organisations and town leaders, to build a collaborative community within a secured and safe environment.

Launched on 13 May 2020 for download on both iOS and Android platform, the application enables residents to identify other like-minded neighbours who have similar passions, allowing them to form interest groups (Kaki Groups), and encourages self-organised initiatives. Redeemable credits (Kaki Points) are also available to Kaki Groups to encourage collective activities through gamification (Figure 9.3).

i) Browse latest happenings
By proximity

ii) Filter happenings according
to specific needs

iii) Respond to happenings: RSVP
or join the Discussion Board

iv) Get real
time-notifications

v) Join or start a Kaki Group &
earn Kaki Points together

vi) Post happenings as an
individual or as a Kaki Group

vii) Compliment a
neighbour

**FIGURE 9.3** Main features of nukampung app shown on user interface designs.

*Source*: Authors.

Operationally, the nukampung app applied strategies mined from case studies in various ways. Besides being a centralised platform with content specially curated for the neighbourhood, the application's announcement and meeting platforms such as 'Town Hall', 'Pasar', 'Just Kaypo',[4] and 'Finding Kaki'[5] are intuitively designed to encourage residents to easily post thoughts on events, sharing of resources, requesting for help, or simply just to find fellow hobbyists through an open call system using standardised interfaces such as 'Kaki Group', 'RSVP', or 'Discussion Board'. Users are also able to browse events happening by mapped priority and proximity by filtering activities, toggling between Constituency, Town, and the rest of Singapore.

Likewise, the idea of gamification from the O2O experience framework inspired the inclusion of fun and user-friendly features such as the reward system of 'Start a Kaki Group', whereby users can initiate their own interest group and gather like-minded users on the nukampung app to host activities and earn 'Kaki Points' together. One can also choose to join an existing group. The underlying motive is to unite pools of residents within the neighbourhood through common interests, with the long-term goal of forming self-sustainable, albeit smaller, but more intimate communities within the larger neighbourhood, that are able to provide not only companionship, but help and support as well when needed. There are hopes that this method of virtual befriending can act as an ice-breaker, scaffolding those who are more reserved in nature, or lack time to approach fellow enthusiasts in person.

Learning from common online behaviours, such as the penchant for individuals to come together over common topics and share personal experiences or comments, the nukampung app encompasses various sharing platforms, each with a specific purpose. 'Just Kaypo' is for the sharing of content in the form of recommendations, tips, or announcements. In 'Discussion Board', users are able to converse with each other through leaving comments or accepting invitations to events. Under 'Host an Event', one can organise virtual or even in-person activities to share skill sets or offer expertise.

The nukampung app was first introduced via community outreach to residents of three towns. The pilot testing of the application was planned to take place complementarily to the physical community engagements organised as part of the research study, in an attempt to observe the uptake of the application in relation to its introduction at the physical activities taking place onsite.

In preparation for the app launch, marketing collaterals and strategies were set up to publicise the app. Due to COVID-19 restrictions in 2020, the engagements became hybridised, with some converted to online discussions. For example, a series of E-Happenings for Project Zero (Chapter 8) in Jurong East were conducted over virtual meetings during the Circuit Breaker period. These sessions were facilitated by the research team, where a segment was set aside for the sharing of the nukampung app through demonstrations and guided installation with the nukampung technical team. When physical restrictions eased, the research team guided the participants to envision starting their own initiatives using the 'Kaki Group' function in a physical Chit-Chat session.

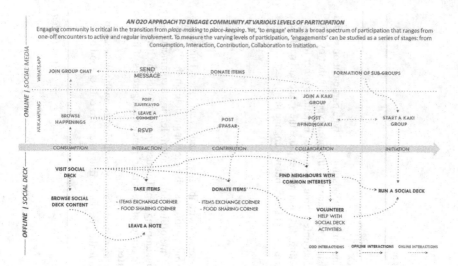

**FIGURE 9.4**  O2O Scenario mapping.

*Source*: Authors.

During the physical prototyping phase of Social Deck (Chapter 8), the nukampung app was the main communication tool facilitating residents in their management of self-initiated community activities. By January 2021, COVID-19 restrictions were relaxed further, allowing up to eight people to gather. This inspired the 'Power of 8' Campaign, which was launched at the opening event of a National Kidney Foundation centre in Jurong East, where the research team took the opportunity to approach residents to form 'Kaki Groups' of eight persons and participate in the challenge. The initial period of the challenge ran until end-February, with a subsequent one until end-April.

Figure 9.4 shows the usage of the nukampung app in relation to the O2O experience framework demonstrating the online-to-offline and offline-to-online transition of activities at each stage. Each level of participation relates to how the nukampung app is being utilised, where user actions are executed via specific touchpoints, whether through the nukampung app, other online chat groups, or physical interactions at Social Deck. Detailed descriptions of activities are outlined in Table 9.2.

## Data Analysis and Evaluation

To gain a contextualised understanding of the adoption and operation of the nukampung app, the number of user sign-ups, posts, and Kaki Groups were monitored over a period of a year, from May 2020 to June 2021 (Table 9.3).

Jurong East has the most number of users (261), followed by Punggol (83), and, lastly, Toa Payoh (21). Jurong East also has the most number of posts and Kaki Groups formed. Using Jurong East as an example, the sign-up trend of the

TABLE 9.2 Online-to-offline/offline-to-online transition of activities at each stage of user experience

| Participation | Online | Offline |
|---|---|---|
| **Consumption** | **nukampung**<br>Browse happenings: The user scrolls through the nukampung home page to browse happenings in their constituency, town, and the rest of Singapore. They may filter through any of the given categories – (1) Just Kaypo, (2) Finding Kaki, (3) Events, (4) Pasar, and (5) Town Hall – to narrow their search.<br>As the user browses the home feed, they might select posts of interest and get redirected to third party links outside of nukampung. This denotes a passive, one-way type of participation | **Social Deck**<br>In the offline context, 'Consumption' type participation would entail visiting the place-making prototype, whether by chance encounter, through word of mouth, or other forms of marketing. The user may explore the space and read the contents at Social Deck, but not engage in any activities |
| **Interaction** | **nukampung**<br>Discussion Board: The user engages with other nukampung users and the content found on nukampung by engaging in conversation via the discussion board – leaving a comment or indicating attendance through the 'RSVP' button | **Social Deck**<br>The user interacts with the space and its activities. This may be done through actions such as leaving a note on the Post-Its provided, the offline equivalent of 'leaving a comment' on the nukampung app. The user may also visit the items exchange corner at the 'Pasar' panel, taking items of interest<br>**Self-initiated**<br>The user may connect directly with donors on the WhatsApp group to make personal arrangements to collect food or items |

*(Continued)*

**TABLE 9.2** (Continued)

| Participation | Online | Offline |
|---|---|---|
| **Contribution** | **nukampung**<br><br>Just Kaypo (Post happening): The user contributes to the nukampung community by sharing content in the form of recommendations, tips, or announcements<br><br>Pasar (Post happening): The user shares resources on nukampung. This may be surplus items or food that they are looking to give away to others | **Social Deck**<br>The user goes beyond engaging with the activities of Social Deck, but supports the initiative by donating items or food or helping to organise and manage the space. This reflects a high level of ownership of the space where the individual has agency to upkeep the space and further promote it<br>**Self-initiated**<br>The user may connect directly with those who request assistance on the WhatsApp group |
| **Collaboration** | **nukampung**<br><br>Find Kakis (Post happening): The user posts a 'Finding Kaki', seeking collaboration by identifying and activating others in the nukampung community – whether to join a new Kaki Group, volunteer for a cause, carpool, or provide ad hoc assistance<br><br>Join a Kaki Group: The user can browse existing interest groups on nukampung and request to join the community. Groups can earn Kaki Points together to participate in periodic challenges hosted by the nukampung admin | **Social Deck**<br>The user may support the initiative by collaborating with the Community Champions to host events and volunteer on a regular basis<br>**Self-initiated**<br>The user may collaborate with others, whether connected through the WhatsApp group or nukampung, to be engaged in activities unassociated with Social Deck |
| **Initiation** | **nukampung**<br><br>Host an event (Post happening): The user initiates an event open to the nukampung community. This can be in the form of a virtual event or physical event in the constituency of choice<br><br>Start a Kaki Group: The user can initiate their own interest group and gather like-minded users on nukampung to host activities and earn Kaki Points together | **Social Deck**<br>The Community Champion will host a Social Deck at their location and duration of choice<br>**Self-initiated**<br>The Community Champion may run their activities independent of Social Deck |

*Source*: Authors.

**TABLE 9.3**  Compilation of number of users, posts, and Kaki Groups

| Town | Jurong East | Punggol | Toa Payoh | Total |
|---|---|---|---|---|
| Total users | 261 | 83 | 21 | 365 |
| Total posts | 113 | 2 | 0 | 115 |
| Total Kaki Groups | 19 | 0 | 0 | 19 |

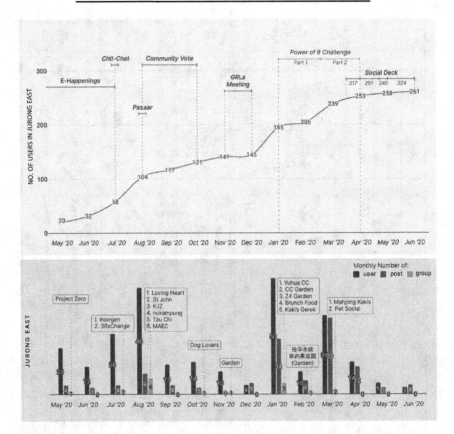

**FIGURE 9.5**   Monthly nukampung data in Jurong East (May 2020–June 2021).

*Source*: Authors.

nukampung app alongside various campaigns and outreach efforts to evaluate their effectiveness is illustrated in Figure 9.5.

A general increasing trend with spikes in user sign-ups during campaigns or engagements is observed. For instance, there were 58 new sign-ups in Jurong East from May to July 2020 when the application was first launched. This coincided with E-Happenings when digital literacy skills were imparted to participants, getting them to become more comfortable with digital tools including the nukampung app to post their events, and kick-starting their first round of contents.

The Chit-Chat session in July 2020 led to the ideation of four Kaki Groups: 3RxChange, Intergen Buddy, MAEC Yuhua, and Kampung Jurong Zine, the first two of which were created on the nukampung app. In line with the launch of the Community Voting Mural Installation in August 2020, five more Kaki Groups were created on the nukampung app – MAEC Yuhua, Tzu Chi SEEN, Kampung Jurong Zine, St John Singapore, and Friends of Loving Heart (Figure 9.5).

The cumulative impact of the Chit-Chat session, together with other events such as the Community Voting Mural Installation that ran from August to October 2020, and the Pazaar held in August 2020, produced a second peak with 46 new user sign-ups, 9 new posts, and 7 new Kaki Groups. The new Kaki Groups were also shared across other social media platforms, reaching over 9,000 people on Facebook, with 926 post clicks and 39 reactions recorded.

The prolonged COVID-19 restrictions stunted the launch of the physical place-making prototypes, resulting in having to rely on the nukampung app and other digital platforms more than anticipated. The Power of 8 Campaign was launched in two phases, Phase 1 from 9 January to 28 February 2021 and Phase 2 from 1 March to 30 April 2021, which correspond to the two spikes in number of users, posts, and Kaki Groups as observed in Figure 9.5. An additional six Kaki Groups were created in Yuhua upon the commencement of the Power of 8 Challenge: Kaki's Gerek, Brunch Food, Yuhua CC, Yuhua CC Garden, Zone 4 Garden, and Yuhua Herb Garden, followed by another two, Mahjong Kakis and Pet Social, in the following phase. While there is increased activity observed during the same week as the engagement events, this usually plateaus after the events are over. Notably, some of the new online connections gradually progressed to direct communications between individuals using other digital communication platforms that they were already familiar with.

Based on a post-implementation review, the significant spikes in user sign-ups during campaigns or engagements in Jurong East as compared to the other two towns were due to: (1) the relatively stronger support from the grassroots leaders in Jurong East; and (2) the active ground engagements associated with the implementation of place-making and place-keeping activities like Social Deck. With the support from grassroots leaders, a series of community engagements leading up to the app launch also led to more initial sign-ups in Jurong East and Punggol. Despite the research team's as well as the collaborating authorities' extensive outreach efforts in Toa Payoh, the lack of grassroots' leaders' support led to a generally lukewarm reception. Furthermore, the mature demographic of Toa Payoh, consisting of a large proportion of Resident Archetype 1 Golden Seniors and Resident Archetype 5 Silver Contributors, with many being Chinese or dialect speaking elderly (Chapter 2), presented a barrier in terms of digital and language literacy as the pilot app was only available in English.

Feedback from the community and stakeholders provided constructive information pivotal to the improvement of the nukampung app. Concerns regarding the app's security were voiced by some grassroots leaders. To address this, discussions ensued with the nukampung app technical team to assess existing security measures and formulate a future plan for robust mobile app security. Measures include secure authentication, backend monitoring, spam management, collaborative community moderation, and the introduction of 'Public Kaki Groups'. Recommendations for the future involve incorporating Artificial Intelligence (AI) for content identification and increasing manpower for enhanced content moderation.

Grassroots leaders in Yuhua Constituency revealed positive sentiments towards the nukampung app, suggesting potential adoption by urban farming and agritech groups in Yuhua, endorsing its publicity at Yuhua Community Centre. In response, the research team swiftly collaborated with urban farming group leaders in Yuhua, organising app-sharing and installation sessions in January 2021. Extensive marketing collateral, including posters, flyers, and banners, adorned lift lobbies in Yuhua. An additional opportunity arose when the research team was invited to contribute information panels for the National Kidney Foundation (NKF) opening event in January 2021, serving as a soft launch for Social Deck and the nukampung app as complementary features for community development.

In Punggol Shore Constituency, the nukampung app received favourable feedback; this positive response prompted the research team to initiate app challenge publicity in early January 2021. The soft launch of the nukampung app with Cascadia Resident Network prompted subsequent discussions with residents that led to the strategic implementation of constituency-based grouping of users, highlighting constituency zone-based activities for easy search results, as depicted in Figure 9.6.

## Conclusion

The integration of digital platforms such as nukampung app, with physical community initiatives, has exemplified that there is significant potential to seamlessly bridge the gap between online and offline realms. This novel approach to community building enables activities to toggle between both realms effortlessly, widening the opportunities for more communities to form and continue. As seen from the transition of activities between nukampung and Social Deck, there is also the potential for O2O strategies to couple with and complement physical infrastructure and activities in place-making and place-keeping. Nevertheless, it is essential to note that success of the implementation of strategies is dependent on the robust support from the ground, e.g. grassroots leaders, authorities, local stakeholders, social service providers, etc. This aligns with the Community Enablement Framework's (Chapter 8) emphasis on matchmaking and integrating into the local ecosystem, necessitating adaptations and customisations to meet specific community needs. Additionally, the importance of implementing stringent security measures cannot be overstated, especially in areas with a highly educated population like Punggol, where concerns about privacy and data security contributed to a lower adoption rate compared to regions like Jurong East.

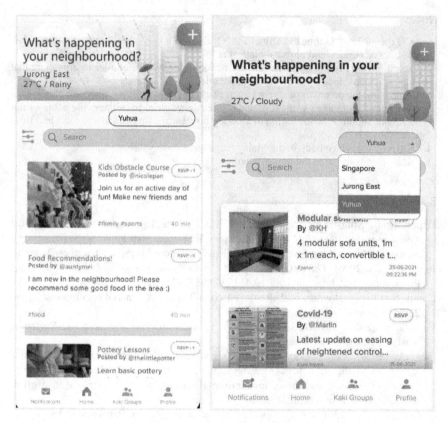

**FIGURE 9.6**  Original nukampung homepage (left), updated nukampung homepage with the addition of the constituency zone toggle (right).

*Source*: Authors.

As we move from the COVID-19 restrictions to a post-COVID era, the potential for scaling these initiatives grows, shifting from a focus on immediate needs to broader interests or advocacy. This transition is evidenced by the emergence of new applications and community platforms post the pandemic like Good Hood (which nukampung has handed over to after the research ended) and the SG Assist App, indicating a potentially sustainable model for O2O community engagement in the evolving social landscape.

## Acknowledgement

The authors would like to thank Professor Yuen Chau and Researcher Iresha Mihirani, formerly at Singapore University of Technology and Design, for developing the backend system for the nukampung mobile app, as well as Swag Soft Holdings Pte. Ltd. for assisting in the mobile app user interface development.

## Notes

1 O2O means both 'Online to Offline' and 'Offline to Online', the two-way flow of activities between digital and physical platforms.
2 Circuit Breaker refers to the mandatory nationwide stay-home order in Singapore that was exercised due to the COVID-19 pandemic.
3 Composing the acronym of 'New Urban Kampung', as well as a phonetic play on the word 'new', combined with 'Kampung', the Malay word for village, the name of nu-kampung signifies the aspiration of creating novel ways for residents to meet and form community bonds.
4 'Kaypo' means busybody in the Malay language.
5 'Kaki' means friend in the Malay language.

## Bibliography

Deputy Prime Minister (DPM) Heng, S. K. (2019, June 15). *In full: DPM Heng Swee Keat's speech at the REACH-CNA dialogue.* Retrieved from https://www.pmo.gov.sg/Newsroom/DPM-Heng-Swee-Keat-Building-Our-Future-Singapore-Together-Dialogue

Jupp, V. (Ed.) (2006). *The SAGE dictionary of social research methods (Vols.* 1-0*).* Sage Publications, Ltd. https://doi.org/10.4135/9780857020116

Kemp, S. (2020). *Digital 2020.* Retrieved from https://datareportal.com/reports/digital-2020-singapore

Low, G. S. (2020). COVID-19: Here's how you can support your favourite Singapore hawker stalls. *Channel News Asia, CNA.*

Ong, A. (2020). Commentary: COVID-19 has revealed a new disadvantaged group among us – digital outcasts. *Channel News Asia, CNA.* Retrieved from https://www.channelnewsasia.com/news/commentary/covid-19-has-revealed-digital-divide-literacy-singapore-12783252

Salkind, N. J. (Ed.) (2010). *Encyclopedia of research design.* Sage Publications, Inc. https://doi.org/10.4135/9781412961288

The Infocomm Media Development Authority (IMDA). (2023, November 4). *Singapore Digital Society Report.* The Infocomm Media Development Authority (IMDA). Retrieved from https://www.imda.gov.sg/resources/press-releases-factsheets-and-speeches/press-releases/2023/singapore-digital-society-report

Yip, C., & Smalley, R. (2020). Home-based learning blues: Life in a rental flat during the COVID-19 circuit breaker. *Channel News Asia, CNA.*

# 10

# APPLICATION OF SMART COMMUNITY DESIGN

*Keng Hua Chong, Pia Fricker, Christine Yogiaman and Mihye Cho*

## Reality – Implications on Housing Policy and Neighbourhood Design

The extensive data-driven studies and community prototyping outlined in the preceding chapters have yielded numerous policy and design insights regarding the future of housing development, applicable not only to Singapore but also to other urban centres grappling with similar trends and challenges.

For instance, many developed cities are grappling with the challenge of an ageing population. By 2030, nearly one in four Singaporeans will be 65 years or older, and by 2050 nearly one in three, compared to one in five today. However, addressing the issues associated with an ageing population extends beyond healthcare provision alone. Our research indicates that more than a quarter (26.1%) of emerging resident archetypes comprise seniors living alone or with other seniors, representing a demographic at risk of lacking robust social support networks. To address this challenge, we advocate for augmenting the prevailing concept of "ageing-in-place" with the principle of "ageing-in-community" (Thomas & Blanchard, 2009; Chong & Cho, 2018). This entails providing comprehensive support not only from healthcare professionals but also from social workers, local businesses, interest groups, and grassroots organisations. It necessitates a redefinition and redesign of residential neighbourhoods to offer a broader range of affordable assisted living housing options, complemented by innovative urban spaces and community initiatives aimed at facilitating outdoor mobility, social interaction, and mutual support among this demographic.

In response to this pressing challenge, the Singapore government has embarked on a series of innovative housing initiatives aimed at enhancing the quality of life for seniors and promoting active ageing. One notable development is the introduction of Community Care Apartments within Housing & Development Board (HDB) estates (Channel News Asia, October 27, 2022). These apartments are specifically

DOI: 10.4324/9781003437659-14

designed to cater to the needs of seniors, providing them with a supportive environment with healthcare and social care facilities while enabling them to maintain their independence. Various Community Care Apartments were piloted in Bukit Batok, Queenstown, and Bedok (The Straits Times, March 23, 2023). These initiatives not only offer seniors affordable and accessible housing options but also foster intergenerational bonding by integrating both elder-friendly and child-friendly features into residential developments (The Straits Times, October 30, 2023). By 2030, 30 of such assisted living facilities will be constructed to meet the evolving needs of seniors as they age, while subsidising upgrades to seniors' homes and implementing neighbourhood improvements in areas with significant elderly populations to enhance safety and mobility (Channel News Asia, November 16, 2023). Moreover, a substantial budget of S$3.5 billion has also been allocated to supporting seniors in ageing actively and staying socially connected (The Straits Times, February 19, 2024). These concerted efforts underscore Singapore's proactive approach in housing development to addressing the challenges posed by its rapidly ageing population.

Besides ageing, Singapore's demographic is also increasingly diverse. In the past most residents lived with family in self-owned flats. However, a growing number of residents now do not fit such traditional living arrangements. A broader range of housing options is thus imperative. Our research revealed that almost a quarter (24.4%) of emerging resident archetypes do not own homes and hail from varied backgrounds. This demographic spectrum spans from lower-income families or individuals renting subsidised flats from the government to young parents or single adults with higher incomes and education levels leasing flats from the open market for various reasons. The recent pandemic has accelerated the trend towards rental living as a viable option, particularly among younger individuals seeking space and privacy for remote work or study, while also yearning for community support through co-living arrangements with like-minded peers. Consequently, there is a growing urgency to dissociate rental flats from lower-income groups, not only to destigmatise such accommodations but also to foster more inclusive, mixed-demographic housing typologies that promote mutual support. This calls for the timely exploration of co-living as a prototypical housing concept, reallocating resources towards communal spaces rather than individual apartments, which can be especially advantageous for a land-scarce city-state like Singapore.

In response to these emerging needs, the Singapore government has initiated various measures to study and provide additional rental housing options. One notable initiative involves the HDB piloting a new type of public rental housing tailored for low-income singles, featuring individual rooms with shared facilities, which aims to provide affordable yet comfortable living arrangements while promoting social interaction among residents (The Straits Times, March 13, 2023). Efforts are also under way to ensure more equitable access to housing types for singles (The Straits Times, July 31, 2023). Furthermore, ongoing studies explore strategies to foster social mixing and dispel stereotypes in public rental housing developments (Channel News Asia, July 17, 2023), reflecting the government's commitment to addressing the evolving housing needs of Singapore's diverse population while fostering social inclusion and community cohesion.

However, it is essential to ensure that such new housing policies and developments must complement rather than undermine the national objective of promoting the core family nucleus as the foundation of society, which constitutes the remaining half (49.5%) of the emerging resident archetypes in our research. Like many developed cities, Singapore is currently facing a significant challenge with its record-low fertility rate, falling below 1 for the first time, far below the replacement rate of 2.1 (The Straits Times, December 21, 2023). Apart from existing monetary incentives, the public housing policy can potentially play a pivotal role in reversing this trend. Indeed, if planned and designed thoughtfully, new forms of intergenerational co-housing, complete with appropriate amenities and programmes, could create a more appealing, family-friendly environment. Introducing new housing typologies alongside nature place-making and pedestrian-friendly urban design could also contribute to the development of more child-friendly neighbourhoods. These measures, if implemented well, have the potential not only to support diverse demographics but also to help reverse the trend of an ageing population over the long term. Further studies in this area are warranted to explore and refine these strategies for maximum effectiveness.

When it comes to planning and future-proofing, the data-driven approach and predictive modelling proposed in the various chapters have enabled us to comprehend residential neighbourhoods at a more granular level, moving beyond the traditional macro town classifications. Our research indicates that the conventional classification of towns based on young town, middle-age town, and mature town, one that was used since the 1990s, becomes less pertinent when implementing targeted interventions within a single town. This is evident, for example, when new housing developments emerge within mature towns or when there is a concentration of older flats demanding elder-friendly amenities in younger towns. Consequently, there is a need for a redefined housing classification system.

This call for a new classification system is echoed in recent developments within the HDB. In response to the evolving needs of residents and to ensure greater affordability and fairness in public housing, HDB has introduced a new housing classification system consisting of three categories: Standard, Plus, and Prime (The Straits Times, August 25, 2023). The Standard scheme represents traditional HDB flats, providing essential housing options for Singaporeans. The Plus flats are strategically located in "choicer" areas such as near the MRT station or town centre, enhancing their desirability and value. The Plus scheme offers additional subsidies and benefits, but also comes with tighter conditions. Finally, the Prime scheme provides even greater subsidies and support for first-time homebuyers in desirable locations, usually closer to the city centre, and has the tightest conditions. With this multi-tiered approach, more diverse housing needs can be fulfilled regardless of whether the town is young or old, while ensuring a good social mix and promoting inclusivity within HDB estates.

In terms of place-making, our collaborative prototypes that were developed in partnership with residents have highlighted the inadequacy of the current void deck setup and the necessity for strategic programming and contextual design interventions to foster resident bonding and community engagement. It is essential not only to identify the needs but also to leverage the assets of local residents and collaborate with locally based social or commercial partners to establish and manage innovative, interest-based social spaces and community programmes tailored to local contexts and aspirations. In addition to place-making efforts, our participatory action research highlights the importance of promoting "place-keeping" to empower residents to co-manage these "white spaces" – common areas in new HDB precincts designated for neighbours to co-create social spaces and facilities and make these spaces their own (Housing & Development Board [HDB] website, accessed March 22, 2024). Formalising ownership contracts between self-initiated community groups and the local authority can grant residents a higher level of autonomy while instilling a sense of responsibility. Further studies can be pursued to explore how the proposed community enablement approach (Chapters 8 and 9) enhances community participation, fosters a sense of ownership, promotes social sustainability, and enhances overall well-being.

Finally, our participatory action research has revealed that effective community design and development hinges not only on the initiative of individual champions and robust grassroots networks but also on the crucial facilitation provided by community design professionals. These professionals play a pivotal role in understanding data and leveraging smart technology in their engagement and empowerment work, building residents' capacity, navigating complex administrative processes, fostering collaboration among local stakeholders, translating ideas into tangible place designs, and bridging community spaces and programmes through both online and offline approaches. Expanding upon our proposed community enablement framework, it is imperative to establish and elevate the role of community design professionals, essentially "Smart Community Designers", beyond the current reliance on volunteers and social workers. To this end, a parallel track of community design roles can be cultivated within both the public and private sectors. Moreover, dedicated funding is essential to support these professionals in conducting data analytics, training resident champions, and developing new community initiatives.

By professionalising community design while leveraging on emerging data science and technology to better understand evolving needs of shifting demographics, we can collaboratively plan and design urban spaces and housing neighbourhoods that foster greater inclusivity, strengthen social cohesion, and promote a more resilient, collaborative, and harmonious society. The following studios serve as imaginative design research that showcase the application of such a data-driven Smart Community Design approach – the Urban Studies and Planning Challenge Studio at Aalto University, Finland, and the Housing Design Studio at Singapore University of Technology and Design (SUTD). These cross-cultural experiments, each pushing boundaries in urban planning and architectural design, respectively, further exhibit the universal applicability of the research and design methodology.

## REIMAGINATION – CASE 1: URBAN TRANSFORMATIONS INFORMED BY FLOWS OF INTERACTION

Pia Fricker

### Urban Studies and Planning Challenge Studio

In the face of complex challenges posed by urbanisation and the urgent realities of the climate crisis, there is a need to move beyond traditional solution-driven approaches in urban planning and design. These challenges, which encompass environmental, political, and societal dimensions, including socio-economic segregation, highlight the multifaceted nature of urban dynamics (Burry, 2020). The intricate interplay between human and non-human elements within urban environments not only complicates these dynamics but also unveils unique opportunities to craft innovative urban typologies. This perspective, recognising the intricate interplay between human and non-human elements within urban environments, not only highlights the complexities of urban dynamics but also identifies opportunities for innovative urban typologies, where functions emerge organically, fostering bottom-up initiatives (Sayegh, 2020). Building on this, the focus shifts from relying on predefined solutions within existing domain-specific solution-spaces to actively constructing permeable spaces that accommodate such emergent methodologies within the urban framework, emphasising the importance of flexibility and adaptability in urban planning (Fricker, 2022). This entails fostering adaptive and resilient urban systems that guide processes rather than imposing control strategies, with careful consideration given to the diverse needs and interactions of both human and non-human inhabitants (Kirwan & Zhiyong, 2020a).

Within this paradigm, this research, conducted within the framework of the master-level Urban Studies and Planning Challenge Studio 2 at Aalto University, Finland, emphasises a shift towards inclusive problem formulation and exploration (Fricker et al., 2020; Fricker & Girot, 2013). Through collaborative inquiry and transdisciplinary discourse, experts from diverse domains collectively redefine and reframe the fundamental questions at hand, paving the way for nuanced and contextually informed insights into current and future urban challenges, supported by novel data-informed design and planning processes (Bhooshan & Vazquez, 2020; Marvin et al., 2015).

## *Unveiling Emerging* Urban Dynamics by Exploring Complex Systems

This speculative, multi-perspective investigation builds upon theories rooted in complex systems, including self-organisation, emergence, and feedback loops, to analyse and interact with existing data across various scales (Kirwan & Zhiyong, 2020b). It extends these foundational concepts to incorporate diverse generative artificial intelligence (AI) methods (Yang et al., 2024), enhancing our understanding and predictive capabilities within urban dynamics (Forrester, 1969). These theories offer insights into the intricate dynamics of urban systems, enabling us to unveil latent patterns and comprehend the nuanced interactions within them. Specifically, our examination extends to elucidating how self-organisation processes engender the emergence of urban structures, unveiling the underlying mechanisms that drive or prevent urban development. In examining the temporal aspects of interactions, the developed methodology seeks to systematically analyse dynamic feedback loops, utilising them as active components for guiding urban planning and design processes over time (Constanza & Maxwell, 1994; Dodds, 2009). This temporal perspective facilitates a comprehensive understanding of the intricate interplay between feedback mechanisms and urban dynamics, allowing for the discernment of both immediate impacts and long-term implications for urban resilience and sustainability (Yao & Fricker, 2021). By simulating the temporal evolution of these feedback mechanisms, the effectiveness in shaping urban environments is tested and their cascading effects on future urban trajectories is simulated.

## Comparative Study of Punggol and Jurong, Singapore

Two recently developed urban areas in Singapore, Punggol (digital district) and Jurong (lake district), serve as case studies within this research. Punggol has undergone rapid urbanisation and transformation in recent years. Originally a rural landscape, Punggol has evolved into a vibrant residential and recreational hub, characterised by its modern infrastructure and innovative urban design. In contrast, Jurong is undergoing significant redevelopment efforts aimed at enhancing its liveability and sustainability. As one of the oldest industrial areas in Singapore, Jurong is undergoing a transformation to diversify its economy and improve its urban environment.

The comparative study of Punggol and Jurong provides critical insights into the spectrum of challenges and opportunities present in urban development. This research engages with granular local data and distinct situational challenges of these areas (Figures 10.1 and 10.2), aiming to delineate interaction zones and envision potential futures that lay the groundwork for collective discourse (Fricker, 2018). The studio acts as a probing environment, advancing our comprehension of the potential and limitations of current planning and design tools when applied to dynamic urban patterns. In conceptualising the city as an organic entity, it becomes imperative to root our strategies in the local context while acknowledging global environmental constraints. This approach underpins the development of interventions that are sensitive to the unique characteristics of each area, ensuring that urban transformation is both sustainable and grounded in the reality of the place.

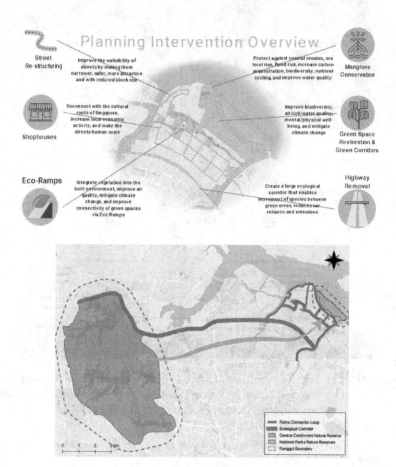

**FIGURE 10.1**  Top: Planning interventions of "Punggulum" project showing post-humanist principles integrated into urban planning and fostering biodiversity; Bottom: Plan showing green network and corridor connecting Punggol's green spaces to other green areas in Singapore.

Student project by Gillian Henderson, Aaron Plaisted, Sameli Sivonen, Jesper Winogradow Image courtesy by Aalto University, Prof. Dr. Pia Fricker.

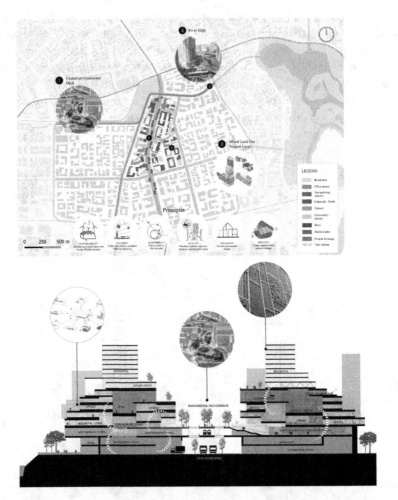

**FIGURE 10.2** Top: The project "Towards Urban Consonance: Envisioning Urban Symbiosis in Jurong" showing multi-dimensional land-use optimisation; Bottom: Section showing the environmental deck as a green artery.

Student project by Roban Colyer, Inari Gustafsson, Julita Koski, Gideon Oladosu, and Maribel Salazar.

Image courtesy by Aalto University, Prof. Dr. Pia Fricker.

The projects featured here are exemplars of what might be achieved when such a nuanced understanding informs urban planning. They stand as references, illustrating a range of possibilities for urban evolution. These case studies underscore the importance of process over prescription, emphasising that the methodologies and outcomes presented here are the result of an iterative, context-aware exploration. This academic work invites stakeholders to a participatory evaluation of future urban development paths, enhancing the shared knowledge base necessary for informed decision-making and the co-creation of future urban landscapes.

The showcased "Punggulum" project, spanning three planning phases (Phase 1: 2025–2030, Phase 2: 2030–2035, Phase 3: 2035–2040), aims for ecological and social sustainability by integrating post-humanist principles into urban planning and fostering biodiversity. The envisioned future includes multifunctional buildings, walkability, and inclusive public spaces (see Figure 10.1). Proposed interventions involve establishing a green network and corridor connecting Punggol's green spaces to larger areas in Singapore and the Central Catchment Nature Reserve. Acknowledging Singapore's unique social landscape, the project addresses challenges of social integration, cultural empowerment, and the ageing population. Efforts to mitigate social disparities through inclusive public spaces promote social cohesion and equity within Punggol. Novel data-analytics support the creation of visionary scenarios through data-informed evaluation and forecasting principles.

The speculative project "Towards Urban Consonance: Envisioning Urban Symbiosis in Jurong" presents an innovative approach to urban redevelopment, grounded in the integration of climate, social, and environmental dynamics. Addressing the need for multi-dimensional land-use optimisation in a city-state with limited space, the master plan (see Figure 10.2, top) dissolves traditional boundaries between residential and industrial zones. It pivots towards a model where mixed land-use and eco-industrial principles align to forge sustainable urban ecosystems. Strategies for mitigating industrial pollution, enhancing spatial efficiency, and fostering community integration are deeply embedded in the fabric of the proposal. By conceiving uninterrupted flows of people, data, and ecology as the backbone of the plan, the project weaves a cohesive tapestry where industrial and residential spheres not only coexist but flourish together. The environmental deck, detailed in Figure 10.2 (bottom), becomes a vital green artery, introducing biodiversity and accessible public spaces into the urban milieu. This transformation champions a virtuous circle of urban regeneration, advocating for high-density living that is resilient and attuned to the challenges of Singapore's hot-humid climate.

## REIMAGINATION – CASE 2: SUPERDIVERSITY, URBAN LIVING FOR ALL

Christine Yogiaman

### Housing Design Studio

Housing Design Studio is the third course in a three-part Fundamentals in Design Studio series offered in the undergraduate Architecture programme in SUTD. Intended to be the terminal design foundation training for undergraduate students, the course probed into the complexities of housing through the various learnt technical and theoretical capacities the students had built up. The studio brief challenged current dominant understanding of housing defined through categorisation of the ideal family unit, income, ethnic, racial, immigration status, etc. It asked students to embrace superdiversity that emerged from the socio-economic and environmental-urban conditions arising from constant and rapidly evolving circumstances. The studio aimed to imagine new forms of heterogeneity in housing that tempered with dominant levelling influences, to give forms to the instabilities of urban living in Singapore, and to achieve "Super-Diversity and Urban Living for All".

### Learning through Social, Urban, and Environmental Goals and Datascape

With much of its economic policy tied to population growth, Singapore's current population density of 8,358 person/km$^2$ will continue its steady increase. Already, the current gross plot ratio in public housing development has climbed beyond 4, with the majority of this development manifesting in an ever-increasing contest of building height. It became imperative that a Housing Design Studio sited in Singapore had to reconsider new building typologies in this eminent future of increased density. The studio began with the immersion into a diverse collection of existing housing building types to uncover the urban, social habitation, and environmental contexts that formulated these formal categories. The studio posited that the study of types allowed students to transcend strict formal classifications and interpret change and adaptation necessary in the constantly evolving multiple realities of urban habitation. Informed by the review of contemporary housing challenges and tendencies identified by Oliver Heckman in his introduction to the book *Floor Plan Manual Housing* (2017), the studios were asked to articulate how effective design strategies in the form of building typologies were deployed to respond to the socio-economic, environmental, and urban data and goals of the precedents. A diagrammed summary of the study displayed a non-direct correlation between building typologies and the socio-economic issues they aim to address, suggesting that multiple design strategies and formal qualities were activated to respond to similar contemporary urban, environmental, and socio-economic trends and challenges. These learnt housing typologies springboard the students' ability to manipulate geometric forms and building volume to respond to their proposed housing goals.

## *Defining Public* through Social-Ecological Network Diagrams

Superdiversity sponsors new forms of public life. Students were asked to study and expand on their definition of Public through the drawing of a social-eco-logical network diagram. The studio postulated that urban/social and ecologi-cal/environmental systems could find alignments, to augment the deficiencies of the other and to opportunistically thrive together. In outlining the various urban/social and ecological/environmental data-driven processes to be stud-ied, students would be able to identify opportunities and mutual interdepend-encies that allow the beginnings in the formulation of an urban/social and ecological/environmental system that defines the project narrative. A robust system would allow for self-regulation, adapting to new equilibriums when confronted with unpredictable changes. Equipped with a refined understand-ing of the public, students endeavoured to initiate their proposed urban infra-structure, building volume and unit plants that would empower it. Students would experiment on their understanding of early study on housing typol-ogy, appropriating established building types into hybrids that would be best manipulated to meet the diverse and often contradictory needs of the site, urban/social, and ecological/environmental narratives. Their application of the social-ecological network diagram for a housing proposal is demonstrated in the two projects mentioned below.

The project "Designing for the Third Age" aims to sustain the cycle of vital-ity among the elder population (Figure 10.3). The social-ecological network diagram identifies the interdependent positive health effects of human-nature interactions that contribute to this goal, which translates into a design collage. The project's massing strategy involves weaving together a busy urban con-text, forests, swampland, and a peaceful waterfront. Its design invites nature and activity into its half-open courtyards, peeling in various directions to create open plazas to keep its residents immersed in communal activity.

The project "Community in Circularity" envisions to unite urban communi-ties towards incorporating circular practices into daily life (Figure 10.4). The social-ecological network diagram strives to illustrate a circular, harmonious

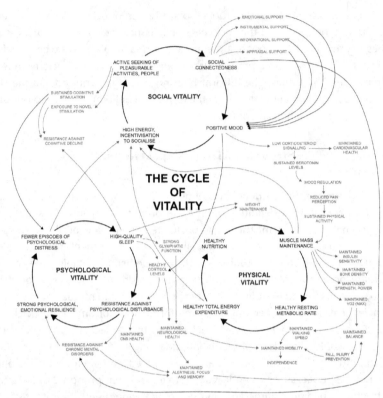

**FIGURE 10.3**   Top: Social-ecological network diagram for a housing proposal "Designing for the Third Age". Bottom: Massing strategy showing urban and nature spaces weaved together (Gross Floor Area: 147,000 m², Plot Ratio: 4.0, Site Coverage: 26%).

Student project by Yeo Soon Yii and Riccia Lim.

Image courtesy by SUTD, Prof. Christine Yogiaman.

**Material
Flow**

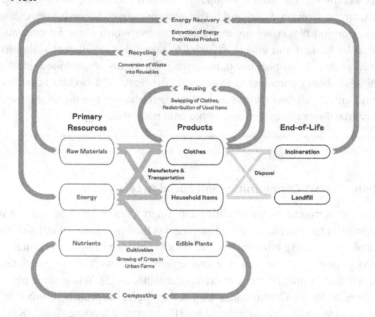

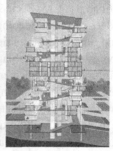

**FIGURE 10.4** Left: Social-ecological network diagram for a housing proposal "Community in Circularity" (top); housing blocks are organised into clusters which allow for communities to be formed and enable activities to take place at different scales (bottom). Right: Section showing residents partaking in activities in this new form of compact living (Gross Floor Area: 287,089 m², Plot Ratio: 5.6, Site Coverage: 43%).

Student project by Aretha Low and Ryan Low.

Image courtesy by SUTD, Prof. Christine Yogiaman.

balance of social, financial, and environmental sustainability, which translates into massing study. Dwelling units are optimised for efficiency and intertwined wAith vibrant shared spaces, including communal kitchens, inviting living spaces, urban farming plots, and activity rooms catered towards various interests. Housing blocks are organised into clusters which allow for communities to be formed and enable activities to take place at different scales. In this new reality of compact living, residents enjoy an uncompromised quality of life while being empowered to actively participate in a circular economy through fostering urban agriculture and resource-sharing among neighbours, and maintaining a symbiotic relationship with natural ecology.

## Reflection – Smart Communities and Social Acceleration

The discourse surrounding Smart Cities and Smart Communities amidst the present geopolitical tensions in the world emphasises the importance of utilising technology and coordinating huge amounts of data to enhance urban management and connectivity, providing targeted and convenient infrastructure and services for citizens, whether in time of peace or crisis (Carvalho, 2015). While some perceive Smart Cities or Smart Communities as a technological approach to urban management undertaken by private companies (Brenner & Theodore, 2002; Kitchin, 2015; Hollands, 2015; Peck & Tickell, 2002), others view them as interconnected assemblages of actors, ideologies, and technologies, whereby the making of Smart Cities and Smart Communities is indeed a political process that intervenes in the formation of these assemblages (Shelton et al., 2015). Despite ongoing debates over definitions and implementation methods, the concept of Smart Cities has been integrated into urban planning globally, including in Singapore, where the Smart Nation vision is a key aspect of national development. There is no doubt that the concept of Smart Communities that leverages both big data and social capital to create social changes would inevitably follow.

However, within the framework of Smart Community Design as elaborated in previous chapters, there is another dimension that affects citizens' quality of life which so far eludes our discussion, i.e. the dimension of time. Different from the modelling of the virtual population where temporal dimension is used in the projection of future, we refer to time here in relation to individuals' experience and perception. In fact, the issue of time came up in many of our interviews, and in some instances even took precedence over physical space, with individuals facing time scarcity ("no time" being the most common reason for not engaging in the community) and prioritising activities such as family and health (by sacrificing one's "free time"). Residents in public housing estates in Singapore experience time constraints, leading to challenges in balancing work and personal life. The notion of "buying time" underscores the perception of time as a limited commodity that can be compensated for with financial resources. This time scarcity affects various aspects of daily life, from family interactions to leisure activities, prompting individuals to seek solutions that maximise time efficiency.

Moreover, the advent of technological advancements, including especially the latest development in AI, contributes to what Hartmut Rosa terms "social acceleration" (Rosa, 2003; Rosa, Dörre & Lessenich, 2017). Under this concept, three distinct spheres –"technological acceleration", "acceleration of social change", and "acceleration of the pace of life" – play pivotal roles in shaping individuals' experiences of time and space within contemporary society.

Technological acceleration refers to intentional speeding up of processes related to transport, communication, and production through advanced technologies, leading to increased efficiency and productivity but also introducing challenges of constant adaptation. The acceleration of social change involves rapid transformation of societal norms, values, and structures, contributing to a reduction of the present moment to reflect, digest, and plan, as well as a sense of instability and flux. Lastly, the acceleration of the pace of life denotes the relentless pursuit of efficiency and productivity in daily activities and routines (evident in the common sight of multitasking), resulting in a constant sense of urgency and busyness.

In light of these spheres, Rosa's analysis highlights how individuals perceive time and space amidst the rapid changes and advancements. The compression of time and space due to technological acceleration makes individuals feel as though the world is moving at an ever-increasing speed, while the acceleration of social change disrupts traditional patterns of behaviour and interaction, challenging individuals to adapt to new norms and expectations. Furthermore, the acceleration of the pace of life leads to heightened stress and anxiety as individuals struggle to find moments of rest and relaxation amidst the demands of contemporary life. Such social acceleration especially resonates with Singapore, which has been defined as a perpetually crisis-ridden country, thereby presenting the adaptability of its citizens as the sole means of survival and national identity, with a sense of urgency to keep up with the pace of change in the technology-driven economy to pursue growth and survival in global competition (Tan, 2012; Yuen, 2018).

As society continues to accelerate, individuals must navigate an increasingly complex and fast-paced world, where the boundaries between work and leisure, past and future, and virtual and physical space become blurred. Achieving a balance between technological advancement and social well-being hence remains a critical challenge, as individuals strive to find meaning and fulfilment amidst the accelerating pace of life (Parkins, 2004; Mayer & Knox, 2009; Pink, 2009; Tambyah & Tan, 2018).

As such, technologists, urban planners, architects, and designers should be highly aware when leveraging technologies and data science to improve quality of life while addressing time poverty and accommodating diverse temporalities. Smart

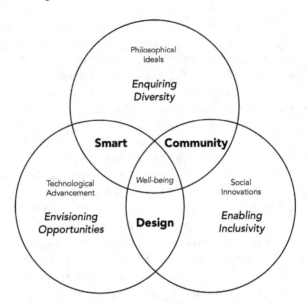

FIGURE 10.5   Smart Community Design framework that brings together philosophical ideals, technological advancement, and social innovations through data-driven participatory process in terms of enquiring diversity, envisioning opportunities, and enabling inclusivity, with the aims to achieve better quality of place, quality of time, and quality of life i.e. well-being. Image credit: Prof. Keng Hua Chong.

Community Design, envisioned as a data-driven participatory process to build social capital and societal changes, could potentially navigate the challenges posed by social acceleration. While spatial interventions may not directly alter working time, they can facilitate better transitions between different aspects of daily life, and between physical and cyber space, recognising the multi-dimensional nature of time and space experienced by individuals. In other words, Smart Community Design can serve as a central ideology and main driver in pursuing better quality of life and well-being for diverse population (Figure 10.5).

By incorporating aspirations for work-life harmony into technological and social innovations, humanising data-driven design through the framework of enquiring diversity, envisioning opportunities, and enabling inclusivity, we believe that Smart Community Design can emerge as field of critical pragmatism that prioritises the well-being of residents, as technological advancements converge with philosophical ideals and societal needs.

## Acknowledgement

### *Aalto University*

We extend our gratitude to the students from the Urban Studies and Planning (USP) year of 2023 for their dedication and creativity, which were essential to the progress of this research. Special thanks are extended to the faculty members from Aalto University, particularly to Christine Mady and Karen Buurmans-Niemi, to Johan Kotze from Helsinki University, and the SUTD for their invaluable support and expertise, which significantly enriched the academic discourse surrounding this work.

### Singapore University of Technology and Design

We extend our gratitude to the students of Housing Design Studio in 2023, the final instalment of the Fundamentals in Design Studio series at the Architecture and Sustainable Design pillar at the SUTD. Their dedication, creativity, and hard work have enriched the studio experience and contributed immensely to the research. We also express our sincere appreciation to our esteemed studio instructors, Calvin Chua, Andrew Lee, and Jackson Tan, for their invaluable guidance, mentorship, and unwavering support throughout the term. Additionally, we extend our thanks to all the guest reviewers who generously shared their expertise and insights during studio reviews, enriching the discourse and fostering a culture of collaborative learning.

### Bibliography

Ahas, R., Järv, O., Saluveer, E., Tiru, M., & Aasa, A. (2015). Everyday space-time geographies: Using mobile phone-based sensor data to monitor urban activity in Harbin, Paris, and Tallinn. *International Journal of Geographical Information Science, 29*(11), 2017–2039.

Angelidou, M. (2015). Smart cities: A conjuncture of four forces. *Cities, 47*, 95–106.

Anthopoulos, L. (2016). Smart utopia vs smart reality: Learning by experience from 10 smart city cases. *Cities, 63*, 128–148.

Barns, S. (2018). Smart cities and urban data platforms: Designing interfaces for smart governance. *City, Culture and Society, 12*, 5–12.

Bauman, Z. (2005). *Work, consumerism and the new poor.* Berkshire; New York: Open University Press; McGraw-Hill Education.

Bhooshan, S., & Vazquez, A. N. (2020). Homes, communities and games: Constructing social agency in our urban futures. *Architectural Design, 90*(3), 60–65. https://doi.org/10.1002/ad.2569

Brenner, N., & Theodore, N. (2002). Cities and the geographies of "actually existing neoliberalism". *Antipode, 34*(3), 349–379.

Burry, M. (2020). Seeking an urban philosophy: Carlo Ratti and the senseable city. *Architectural Design, 90*(3), 32–37. https://doi.org/10.1002/ad.2565

Campbell, A., Converse, P. E., & Rodgers, W. L. (1976). *Quality of American life: The perceptions, evaluations, and satisfactions.* New York: Russell Sage Foundation.

Carvalho, L. (2015). Smart cities from scratch? A socio-technical perspective. *Cambridge Journal of Regions, Economy and Society, 8*(1), 43–60.

Castells, M. (1983). Crisis, planning, and the quality of life: Managing the new historical relationships between space and society. *Environment and Planning D: Society and Space, 1*(1), 3–21.

Centre for Liveable Cities. (2014). *Singapore's evolution into a smart nation*. Singapore: CLC Lecture Series Singapore.

Channel News Asia (CNA). (2022, October 27). What you need to know about the new assisted living HDB flats in Queenstown for seniors. *Channel News Asia (CNA)*. Retrieved from https://www.channelnewsasia.com/singapore/community-care-apartments-hdb-flats-bto-queenstown-seniors-faq-3028446

Channel News Asia (CNA). (2023, January 30). Active ageing centres, employment schemes part of updated plan to help Singaporeans age well. *Channel News Asia (CNA)*. Retrieved from https://www.channelnewsasia.com/singapore/active-ageing-centres-employment-schemes-help-singaporeans-age-well-moh-ong-ye-kung-3239881?cid=internal_sharetool_iphone_04032023_cna

Channel News Asia (CNA). (2023, March 17). Own room but shared kitchen and toilet: Public rental tenants have mixed reactions to new scheme. *Channel News Asia (CNA)*. Retrieved from https://www.channelnewsasia.com/singapore/public-rental-housing-single-room-shared-facilities-hostel-tenants-3351471?cid=internal_sharetool_iphone_17032023_cna

Channel News Asia (CNA). (2023, July 17). The Big Read: Can public rental and BTO flats co-exist harmoniously in the same HDB block? *Channel News Asia (CAN)*. Retrieved from https://www.channelnewsasia.com/singapore/big-read-public-rental-bto-hdb-flats-social-mixing-stereotypes-3630841?cid=internal_sharetool_iphone_17072023_can

Channel News Asia (CNA). (2023, November 16). 30 Community Care Apartments to be built by 2030 to give seniors more housing options [caneo]. *Channel News Asia (CAN)*. Retrieved from https://www.channelnewsasia.com/watch/30-community-care-apartments-be-built-2030-give-seniors-more-housing-options-video-3926021

Cho, M., et al. (2018). "Small places" of ageing in a high-rise housing neighbourhood. *Journal of Ageing Studies, 47*, 57–65.

Chong, K. H. (2018). Reclamation of urban voids and the return of the "kampung spirit" in Singapore's public housing. In K. Chong, & M. Cho (Eds.), *Creative ageing cities* (pp. 19–46). London & New York: Routledge.

Chong, K. H. & Cho, M., Eds. (2018). *Creative ageing cities: Place design with older people in Asian cities*. London & New York: Routledge.

Chua, B. H. (1997). *Political legitimacy and housing: Stakeholding in Singapore*. London & New York: Routledge.

Cocchia, A. (2014). Smart and digital city: A systematic literature review. In R. P. Dameri & C. Rosenthal-Sabroux (Eds.), *Smart city: How to create public and economic value with high technology in urban space* (pp. 13–43). Cham: Springer.

Constanza, R., & Maxwell, T. (1994). Resolution and predictability: An approach to the scaling problem. *Landscape Ecology, 9*(1), 47–57.

Dodds, W. K. (2009). *Laws, theories, and patterns in ecology*. University of California Press. Retrieved from https://ebookcentral.proquest.com/lib/ethz/detail.action?docID=837312

Forrester, J. W. (1969). Urban dynamics. Cambridge, MA; London: MIT Press and Productivity Press.

Fricker, P. (2018). *The real virtual or the* real real*: Entering mixed reality*. Wichmann Verlag. https://doi.org/10.14627/537642044

Fricker, P. (2022). Computing with nature: Digital design methodologies across scales. In R. Monacella & B. Keane (Eds.), *Designing landscape architectural education: Studio ecologies for unpredictable futures* (p. 438). Routledge, Taylor & Francis Group. Retrieved from https://www.routledge.com/Designing-Landscape-Arch

itectural-Education-Studio-Ecologies-for-Unpredictable/Monacella-Keane/p/book/
9780367703653

Fricker, P., & Girot, C. (2013). How to teach new tools in landscape architecture in the digital overload. In R. Stouffs, & S. Sariyildiz (Eds.), *Computation and Performance – Proceedings of the 31st eCAADe Conference – Volume 2, Faculty of Architecture, Delft University of Technology*, Delft, The Netherlands, 18–20 September 2013 (pp. 545–553). Retrieved from https://papers.cumincad.org/cgi-bin/works/paper/ecaade2013_028

Fricker, P., Kotnik, T., & Borg, K. (2020). Computational design pedagogy for the cognitive age. *Anthropologic: Architecture and Fabrication in the Cognitive Age, 1*, 695–692.

Heckmann, O., & Schneider, F. (2017). *Floor plan manual housing*. Basel: Birkhäuser.

Ho, E. (2017). Smart subjects for a Smart Nation? Governing (smart) mentalities in Singapore. *Urban Studies, 54*(13), 3101–3111

Hollands, R. G. (2015). Critical interventions into the corporate smart city. *Cambridge Journal of Regions, Economy and Society, 8*(1), 61–77.

Housing & Development Board (HDB). (2014). *White spaces*. Housing & Development Board (HDB) website accessed 22 Mar 2024. Retrieved from https://www.hdb.gov.sg/about-us/hdbs-refreshed-roadmap-designing-for-life/ilve-connected/white-spaces

Housing & Development Board (HDB). (2023, August 20). *New plus housing model with more subsidies for more affordable BTO flats in attractive locations*. Housing & Development Board (HDB) website accessed 22 Mar 2024. Retrieved from https://www.hdb.gov.sg/about-us/news-and-publications/press-releases/20082023-New-Plus-housing-model-with-more-subsidies

Kirwan, C., & Zhiyong, F. (2020a). City as living organism. In *Smart Cities and Artificial Intelligence* (pp. 25–45). Elsevier. https://doi.org/10.1016/B978-0-12-817024-3.00002-7

Kirwan, C., & Zhiyong, F. (2020b). *Smart cities and artificial intelligence: Convergent systems for planning, design, and operations* (1st ed.). Elsevier.

Kitchin, R. (2015). Making sense of smart cities: Addressing present shortcomings. *Cambridge Journal of Regions, Economy and Society, 8*(1), 131–136.

Marvin, S., Luque-Ayala, A., McFarlane, C., Luque-Ayala, A., & McFarlane, C. (2015). *Smart Urbanism: Utopian vision or false dawn?* Routledge. https://doi.org/10.4324/9781315730554

Mayer, H., & Knox, P. L. (2009). Pace of life and quality of life: The slow city charter. In M. J. Sirgy, R. Phillips, & D. R. Rahtz (Eds.), *Community quality-of-life indicators: Best cases iii* (pp. 21–40). Dordrecht: Springer.

Parkins, W. (2004). Out of time: Fast subjects and slow living. *Time & Society, 13*(2–3), 363–382.

Peck, J., & Tickell, A. (2002). Neoliberalizing space. *Antipode, 34*(3), 380–404.

Pink, S. (2009). Urban social movements and small places: Slow cities as sites of activism. *City, 13*(4), 451–465.

Rosa, H. (2003). Social acceleration: Ethical and political consequences of a desynchronized high–speed society. *Constellations, 10*, 3–33. https://doi.org/10.1111/1467-8675.00309

Rosa, H., Dörre, K., & Lessenich, S. (2017). Appropriation, activation and acceleration: The escalatory logics of capitalist modernity and the crises of dynamic stabilisation. *Theory, Culture & Society, 34*(1), 53–73.

Sayegh, A. (2020). *Responsive environments: Defining our technologically-mediated relationship with space*. New York; Barcelona: Actar D.

Shelton, T., Zook, M., & Wiig, A. (2015). The 'actually existing smart city. *Cambridge Journal of Regions, Economy and Society, 8*(1), 13–25.

Tambyah, S. K., & Tan, S. J. (2018). *Happiness, wellbeing and society: What matters for Singaporeans.* Abingdon and New York: Routledge.

Tan, K. P. (2012). The ideology of pragmatism: Neo-liberal globalisation and political authoritarianism in Singapore. *Journal of Contemporary Asia, 42*(1), 67–92.

The Straits Times. (2023, March 13). HDB to pilot new type of public rental housing with own room, shared facilities for low-income singles. *The Straits Times.* Retrieved from https://www.straitstimes.com/singapore/housing/hdb-to-pilot-new-type-of-public-rental-housing-with-own-room-shared-facilities-for-low-income-singles

The Straits Times. (2023, March 18). Making more room for renting as a housing option in Singapore. *The Straits Times.* Retrieved from https://www.straitstimes.com/opinion/making-more-room-for-renting-as-a-housing-option

The Straits Times. (2023, March 28). 200 assisted living flats for seniors to be launched in Bedok in 2023. *The Straits Times.* Retrieved from https://www.straitstimes.com/singapore/politics/200-assisted-living-flats-for-seniors-to-be-launched-in-bedok-in-2023

The Straits Times. (2023, April 11). Broaden meritocracy to combat social stratification: President Halimah outlines key priorities for Govt. *The Straits Times.* Retrieved from https://www.straitstimes.com/singapore/politics/broaden-meritocracy-to-combat-social-stratification-president-halimah-outlines-key-priorities-for-govt

The Straits Times. (2023, July 31). More equitable access to housing types for singles being studied: Desmond Lee. *The Straits Times.* Retrieved from https://www.straitstimes.com/singapore/more-equitable-access-to-housing-types-for-singles-being-studied-desmond-lee

The Straits Times. (2023, August 20). The Gist: New classification for flats and temporary safety net for retrenched workers. *The Straits Times.* Retrieved from https://www.straitstimes.com/singapore/politics/the-gist-ndr-2023-relooking-hdb-estate-classification-and-retirement-help-for-young-seniors

The Straits Times. (2023, August 25). NDR 2023: New public housing framework needed to ensure affordability, fairness and good social mix. *The Straits Times.* Retrieved from https://www.straitstimes.com/singapore/politics/ndr-2023-new-public-housing-framework-needed-to-ensure-affordability-fairness-and-good-social-mix

The Straits Times. (2023, August 25). NDR 2023: HDB to launch new PLUS flats at choicer locations with 10-year MOP, stricter conditions. The Straits Times. Retrieved from https://www.straitstimes.com/singapore/politics/ndr-2023-hdb-to-launch-new-plus-flats-at-choicer-locations-with-10-year-mop-stricter-conditions

The Straits Times. (2023, September 18). HDB to launch up to 14,000 two-room flexi BTO flats from 2024 to 2026: Desmond Lee. *The Straits Times.* Retrieved from https://www.straitstimes.com/singapore/politics/hdb-to-launch-up-to-14000-two-room-flexi-bto-flats-from-2024-to-2026-desmond-lee

The Straits Times. (2023, October 30). 'We're happy here': How the elderly to young families are enjoying homes distinctly designed for them. *The Straits Times.* Retrieved from https://www.straitstimes.com/singapore/housing/building-homes-together-elderly-young-families-hdb-flats-everyone

The Straits Times. (2023, November 16). Singapore sets aside $800m to help seniors age well at home, in their communities. *The Straits Times*. Retrieved from https://www.straitstimes.com/singapore/singapore-sets-aside-800m-to-help-seniors-age-well-at-home-in-their-communities

The Straits Times. (2023, November 17). Initiative to create pedestrian-friendly streets to cover all 24 Singapore towns by 2030. The Straits Times. Retrieved from https://www.straitstimes.com/singapore/initiative-to-create-pedestrian-friendly-streets-to-cover-all-24-singapore-towns-by-2030

The Straits Times. (2024, February 19). Budget 2024: $3.5 billion to help seniors age actively, stay socially connected. *The Straits Times*. Retrieved from https://www.straitstimes.com/singapore/budget-2024-35-billion-to-help-seniors-age-actively-stay-socially-connected

Thomas, W. H., & Blanchard, J. M. (2009). Moving beyond place: Aging in community. *Generations, 33*(2), 12–17.

Yang, J., Fricker, P., & Jung, A. (2024). From intangible to tangible: The role of big data and machine learning in walkability studies. *Computers, Environment and Urban Systems, 109*, 102087. https://doi.org/10.1016/j.compenvurbsys.2024.102087

Yao, C., & Fricker, P. (2021). *How to cool down dense urban environments? A discussion on site-specific urban mitigating strategies*. Wichmann Verlag. https://doi.org/10.14627/537705007

Yuen, B. (2018). *Singapore: Smart city, smart state by Kent E. Calder*. Washington, DC: Brookings Institution Press, 2016, pp. 233. Appendix, Notes, Bibliography, Index. *Journal of Southeast Asian Studies, 49*(2), 349–351.

# INDEX

Note: **Bold** page numbers refer to tables; *italic* page numbers refer to figures and page numbers followed by "n" denote endnotes.

Printed in the United States
by Baker & Taylor Publisher Services

Printed in the United States
by Baker & Taylor Publisher Services